An Introduction to Chemical Techniques

Peter Tooley PhD MSc ARIC MIBiol
Head of Department of Chemistry
St. Osyth's College, Clacton

John Murray Fifty Albemarle Street London

To my wife, with affection

Also by Peter Tooley

Chemistry in Industry

High Polymers
Fats, Oils and Waxes
Fuels, Explosives and Dyestuffs
Food and Drugs

Experiments in Applied Chemistry

Author's note

The nomenclature used in this book is that suggested by the Working Party of the Education (Research) Committee of the Association for Science Education. To assist those who may be unfamiliar with the recommended names, an appendix is included at the end of the book which contains an alphabetical list of equivalent common names.

'The Devil may write chemistry textbooks, because every few years the whole thing changes.'

Berzelius, 1831

Printed in Great Britain by Martin's of Berwick

0 7195 2651 5

Preface

A chemistry course stands or falls on the quality of its practical work. In turn, success in the practical field depends a great deal upon the skill of the experimenter, who must not only be familiar with apparatus and techniques but understand the theoretical principles behind their use. Attention to detail, intelligent handling of materials and glassware, and scrupulous cleanliness are also marks of experimental competence. This book is intended as an introduction to the principal techniques likely to be used in experimental organic chemistry in the school laboratory.

The methods available for the preparation and purification of organic materials are described in some detail together with a brief account of relevant theory. Where appropriate the development of these techniques for use on a commercial scale is discussed to show the engineering problems involved in moving from gram to tonne capacities.

At the end of each section a number of simple experiments are suggested to illustrate the principles discussed, making use of a wide range of apparatus on both micro and traditional scale. An attempt has been made to introduce a note of originality in these experiments and to make use of materials familiar to the pupil and readily available in the school situation. Suggestions have also been made for the making up of home-made apparatus bearing in mind the restricted budget of most schools.

It is hoped that this volume will prove of general use to pupils starting practical organic chemistry for the first time, and will be of special value as an introduction to *Experiments in Applied Chemistry*.

Contents

Experiments

3	**Filtration** (pages 11–12)	
3.1	Purification of contaminated aspirin	*hot filtration and Hirsch filter*
3.2	Preparation of triiodomethane on a semi-micro scale	*Willstätter nail filter*
3.3	Isolation of peroxidase from turnips	*Buchner filter*
3.4	Hydrolysis of aspirin	*Hirsch filter*
4	**Centrifugation** (page 15)	
4.1	Emulsion breaking by centrifugation	*laboratory centrifuge*
4.2	Use of centrifugation in sedimentation	*laboratory centrifuge*
4.3	Separation of methyl orange and methylene blue by micro-chromatography	*laboratory centrifuge and micro-adsorption column*
5	**Crystallization** (pages 19–20)	
5.1	Purification of crude N-phenylethanamide	*recrystallization and decolorization*
5.2	Recrystallization of impure 1,3-dinitrobenzene	*recrystallization using a volatile solvent*
5.3	Preparation of osazones from glucose and lactose	*recrystallization on a semi-micro scale*
5.4	Preparation of a 'supersaturated' solution	*crystallization (supersaturation)*
6	**Distillation** (pages 27–9)	
6.1	Recovery of dry cleaning fluid	*simple distillation*
6.2	Comparison of distillation assemblies	*fractional distillation, Dufton column, packed column, pear column*
6.3	Fractional distillation (micro scale)	*micro-distillation unit, 'pot scourer' column*
6.4	Fractionation of crude oil mixture	*fractional distillation at reduced pressure*
6.5	Extraction of sugar from beet	*evaporation at reduced pressure*
6.6	Dehydration of ethanol using an azeotropic technique	*fractional distillation*
6.7	Isolation of naphthalene from adulterated sample (mothball)	*steam distillation, small scale*
6.8	Steam distillation of a mixture of 2-hydroxybenzoic acid and 1,4-dichlorobenzene	*steam distillation, semi-micro scale*
6.9	Steam distillation of a mixture of benzene and technical xylene	*steam distillation, semi-micro scale*
7	**Solvent extraction** (pages 34–5)	
7.1	Comparison of single and multiple extractions	*solvent extraction*
7.2	Estimation of distribution coefficient by a colorimetric method	*solvent extraction*
7.3	Separation of benzenecarboxylic acid from 1,4-dichlorobenzene	*extraction by salt formation*
7.4	Demonstration of the 'salting out' effect	*emulsion breaking by addition of water soluble salts*
7.5	Determination of the oil content of peanuts	*Soxhlet extraction apparatus*
8	**Chromatography** (pages 53–9)	
8.1	Separation of methyl orange and methylene blue and a mixture of 2- and 4-nitrophenylamine	*semi-micro column, wet packing*
8.2	Separation of leaf pigments	*mixed column, dry packing*
8.3	Analysis of black and brown inks	*thin layer chromatography*
8.4	Extraction of fluorescing agents from detergent	*TLC and location by fluorescence*
8.5	Trial separation of malachite green and methylene blue	*wedge and sandwich chromatography*
8.6	Purification of anthracene	*locating agent with column chromatography*
8.7	Identification of food dyes by ascending chromatography	*multiple ascending chromatography*
8.8	Analysis of a mixture of indicators	*multiple descending chromatography*
8.9	Separation of aminoacids in human hair hydrolysate	*two-dimensional separation and locating reagent*
8.10	Analysis of inks	*disc chromatography*
8.11	Purification of crude sugar	*ion exchange chromatography*
8.12	Conversion of trisodium(I) citrate to 2-hydroxypropane-1,2,3-tricarboxylic acid	*ion exchange chromatography*
8.13	Deionization of a solution	
8.14	Examination of cigarette lighter fuel	*gas–liquid chromatography*
8.15	Separation of dyes	*electrophoresis*
8.16	Separation of aminoacids in lemon juice	*two-dimensional separation: electrophoresis and chromatography*
8.17	Desalting skimmed milk by gel filtration	

1 Melting points

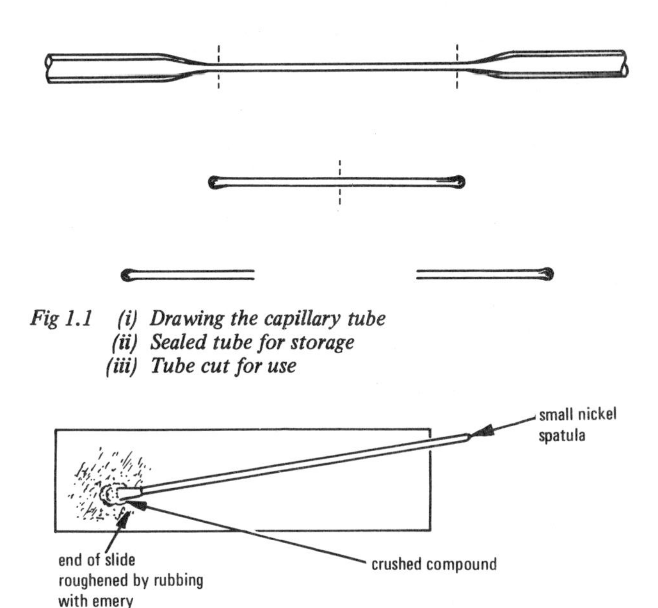

Fig 1.1 (i) Drawing the capillary tube
(ii) Sealed tube for storage
(iii) Tube cut for use

Fig 1.2

A pure crystalline organic solid has a sharp and specific melting point which usually lies within the range 50-300° C. Although melting point is not greatly affected by variations in pressure, the presence of even a small amount of an impurity produces a significant depression. In addition to lowering the melting point the presence of an impurity results in the melting taking place over a much wider temperature range. Occasionally, solid mixtures (eutectics) give precise melting points but these are unusual enough to be disregarded. Thus the accurate determination of melting points affords a simple means of identifying organic compounds and determining their purity.

Sometimes the melting points of two or more organic compounds are very close together or even identical. In this case we can make use of the fact that a mixture of two substances has a melting point which is below that of either of the components. Often this lowering of melting point is quite dramatic, depressions of 15-20 ° C being not uncommon. If the melting point of a known organic substance remains unchanged after mixing with a second substance, then it can be assumed that the two substances are identical. The determination of 'mixed' melting points in this way is a useful analytical technique.

The determination of melting point is carried out by carefully heating a very small sample of the pure dry substance and noting the temperature range over which melting occurs. Ideally this should be less than 1 ° C. The sample is usually enclosed in a thin-walled capillary tube and heated either in a liquid bath, or by an electric heating block. A micro melting point technique can also be used in which a single crystal of the substance in an electrically heated cavity is observed through a microscope.

It sometimes happens that a solid decomposes upon heating, either before or at its melting point. In this case the temperature at which decomposition commences and the melting range are still often of value in identification. A compound may also exist in different crystal forms each having a different melting point.

Experimental details

Thin-walled melting-point tubes, sealed at one end and about 75 mm x 1 mm inside diameter, can be bought quite cheaply. Alternatively they can be made by drawing out a length of 10 mm soft glass tubing which has been softened by heating (fig 1.1 (i)). In the latter case it is a good idea to cut the capillary tube formed into 16 cm lengths which can be sealed at both ends (ii). When required these can then be cut to produce two clean dry tubes (iii).

Before introducing a sample into the melting-point tube it is often necessary to crush the solid to a fine powder. This is conveniently carried out by placing a few crystals of the compound on one end of a microscope slide which has been roughened with emery paper, and gently rubbing with the back of a clean spatula (fig 1.2).

It is not advisable to crush the solid on filter paper because small pieces of paper fluff invariably become mixed with the sample. The open end of the capillary is pressed into the powdered solid until about 2 mm of the capillary is filled. The sample is then conveyed to the sealed end of the tube by tapping the tube sharply against the thumb-nail or by gently drawing a nail file across the open end. With soft waxy substances it may be helpful to drop the tube, sealed end first, down a short length of glass tubing or a condenser held upright on the bench.

Liquid heating baths

When a liquid bath is being used, the lower end of a thermometer of suitable range is wetted by dipping into the bath liquid. The tube is then slid into position along the thermometer stem so that the sample is level with the centre of the thermometer bulb. If the thermometer is now lowered into the heating bath until the tube is half immersed in liquid the latter will remain fixed to the thermometer without the need for a rubber band. The thermometer is positioned by means of a vented cork loosely fitted into the neck of the bath which can be either a boiling tube or Thiele tube (fig 1.3). The advantage of the Thiele tube is that convection currents produced upon heating the elbow of the tube (thermal stirring) make the use of a mechanical stirrer unnecessary.

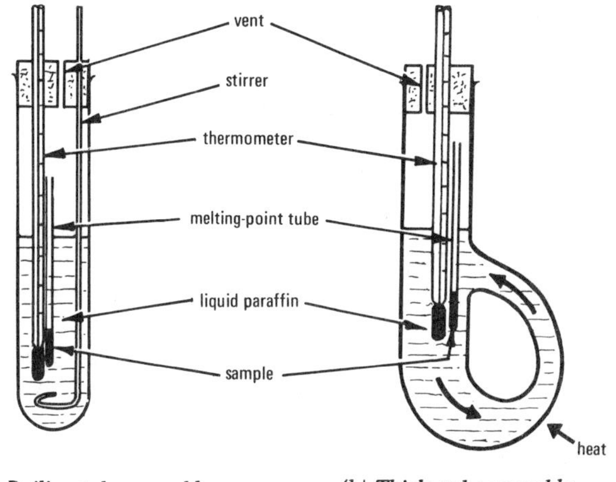

(a) Boiling tube assembly *(b) Thiele tube assembly*

Fig 1.3

The most convenient liquid to use in the heating bath for temperatures up to about 220 °C is medicinal-type paraffin oil. This has the advantages of being cheap, stable and safe to use. For temperatures in excess of this, a medium viscosity silicone oil has a number of advantages, but is rather costly. Fortunately few compounds are encountered with melting points outside the range of a paraffin bath.

It is convenient to prepare two melting-point tubes so that the first can be used for an approximate determination. Once this has been done the bath can be heated quickly to within about 10 °C of the expected melting point and then the temperature raised very slowly with gentle stirring to enable the sample to equilibrate thermally with the bath. A temperature rise of two degrees a minute should enable the melting range to be accurately determined, from the initial softening of the sample until its complete liquefaction. If the melting point is accidentally overshot, no attempt must be made to obtain a result from the temperature at which re-solidification occurs. Neither must the same sample be used twice. Because of the errors which are inevitable in routine laboratory determinations, a temperature difference of about five degrees should be allowed for when using melting point tables for identification purposes.

Electrical heaters

Melting points can be conveniently determined using an electrical heating device (see photograph). This method removes the necessity for a liquid bath and enables high temperatures to be reached quickly and safely. A heater of this type consists typically of an electrically heated metal block containing a horizontal tunnel. The thermometer bulb and the sealed end of the melting-point tube are located in two vertical holes drilled in the block, and lie side by side in the centre of the tunnel. The rear of the tunnel is illuminated with a lamp and the sample can be clearly seen through a lens. The rate of heating can be delicately controlled by means of a rheostat. After use, a water cooled tube can be used to replace the thermometer and reduce the temperature of the block more rapidly. A simple 'do-it-yourself' electrical heating block constructed from an old electric iron has been described by Vogel*.

Electric melting point apparatus

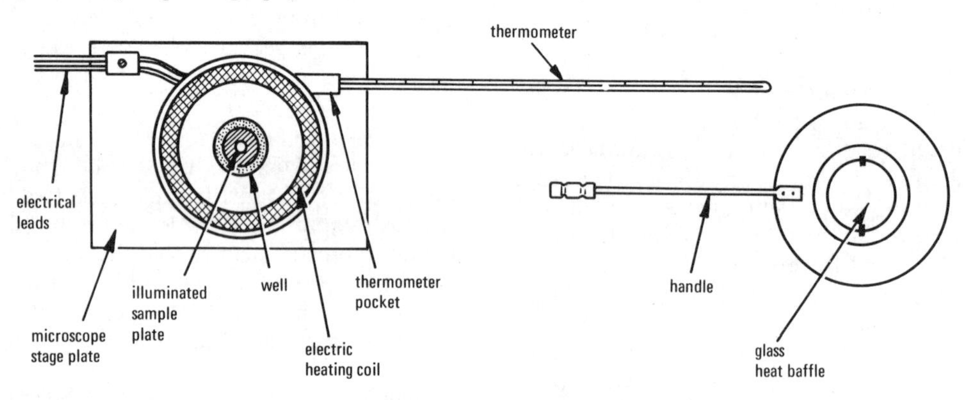

Fig. 1.4 Melting point apparatus (micro scale)

Melting point determinations on a micro scale (fig 1.4) using single crystals can be made with a specially constructed microscope 'hot stage'. This comprises an electrically heated circular hot plate with a central cavity into which protrudes a horizontally set thermometer. The sample is located on a glass slide in the well of the hot plate and covered by a glass heat baffle. The baffle is fitted with a handling stem for convenience. Using the normal optical system of the microscope the hot stage is illuminated from below and the sample observed during heating using a low power objective. The glass baffle prevents thermal damage to the objective lens. A graduated rheostat is used to control the heating of the stage. Occasionally difficulty in using this type of apparatus is experienced as a result of condensation or sublimed solid coating the under surface of the baffle glass.

Sublimation

If a solid has a vapour pressure which exceeds atmospheric pressure before its melting point is reached, then it will pass directly from the solid state to the vapour state on heating, without the appearance of a liquid phase. On cooling, the vapour will condense directly to the corresponding solid. This phenomenon is called sublimation,

***Elementary Practical Organic Chemistry*, A. I. Vogel (Longman, 1960) Part 1, p.79.**

and provides a useful means of purifying a solid, especially on a small scale.

Although the number of substances which can be sublimed at atmospheric pressure is limited, a much wider range of organic compounds sublime at reduced pressures. Small quantities of solid may be purified in this way by gently heating in a filter tube connected to a suction pump. A 'cold-finger' fitted into the tube provides a condensing surface to collect the pure sublimate (fig 1.5).

At atmospheric pressure sublimation can be conveniently carried out by gently warming the impure sample in a small Petri dish covered with a perforated filter paper. As shown in fig 1.6 an inverted funnel with a small plug of glass wool in the stem is placed above the paper. Gentle suction may be applied to the funnel stem but this is not usually necessary. The sublimate is deposited on the surface of the paper and the walls of the funnel. A simpler method is to replace paper and funnel by a large watch glass containing a little crushed ice.

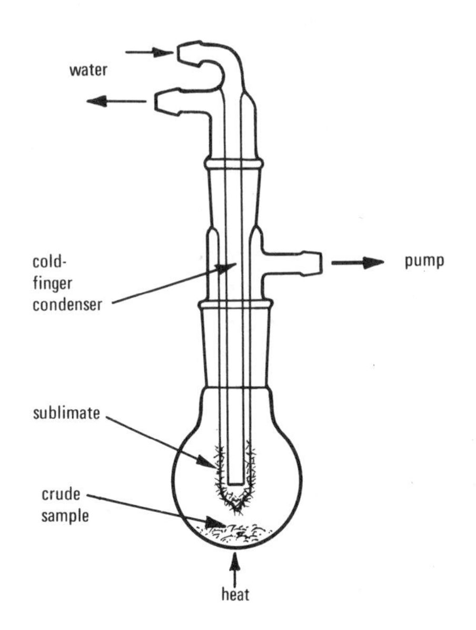

Fig 1.5

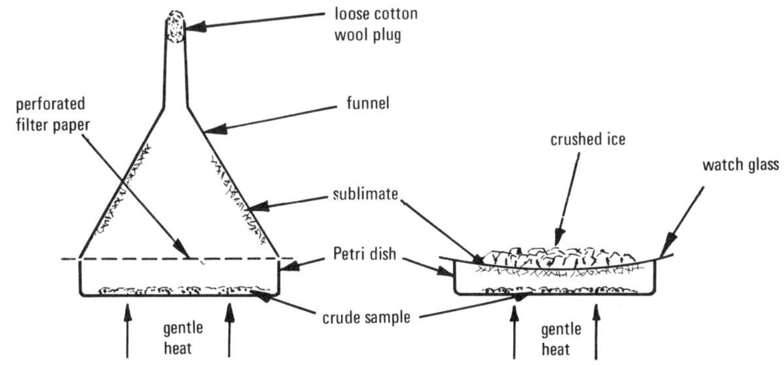

Fig 1.6

Freeze drying

A process known as freeze drying is used to dehydrate heat sensitive materials on a commercial scale by sublimation. In 1909, L. F. Shackell described a method of drying rabies virus by subliming off the ice formed in the frozen material, and removing the resulting water vapour with a chemical desiccant. It was shown that the dehydrated product could be readily reconstituted with water, and retained the properties of the original material.

A quarter of a century later, E. W. Flosdorf used a similar technique for desiccating blood plasma, except that the water vapour was pumped off and removed on a refrigerated condenser plate. Although expensive when compared with other methods of dehydration, freeze drying is ideally suited for the dehydration of materials containing heat sensitive proteins. These include tissue grafts, antibiotics, human milk, certain foodstuffs, and vaccines.

Freeze drying was of great value during the last war in the processing of vast amounts of penicillin and blood plasma. More recently plant has been built in the USA, Italy, France, Britain, and other European countries for the preservation of food by freeze drying.

In a typical dehydration cycle the pre-frozen food is placed on heated trays of expanded aluminium mesh. The trays are then loaded into vacuum tunnels. Careful temperature control is maintained to prevent the ice crystals melting while providing the latent heat for sublimation to occur. The sublimed water vapour is pumped away and removed by condensation on a refrigerated baffle.

Food processed in this way keeps indefinitely at ambient temperatures without refrigeration, while sealed in a plastic pack. Apart from a loss in weight the food remains practically unaltered both physically and chemically.

Melting point experiments

1.1 IDENTIFICATION BY MELTING POINT

(liquid bath apparatus and mixed melting points)

Required

Two sets of semi-micro reagent bottles containing the following powdered substances (melting points in degrees Celsius shown in brackets), one set to be labelled and the other marked A-F only.

Benzene-1,3-diol (110); N-phenylethanamide (115); benzenecarboxylic acid (123); benzenecarboxamide (128); carbamide (132); aspirin (136)

Procedure

1 Choose any two of the unlabelled substances and introduce a sample of each into two melting-point tubes.

2 Determine the melting point of the samples using a Thiele tube for one and a boiling tube and stirrer for the other (fig 1.3).

3 Record the melting point ranges you obtain.

Sequel

Decide which of the compounds listed has a melting point corresponding most nearly to one of your unknowns. Mix three parts of this compound with one part of the unknown and redetermine the melting point. Comment on the result.

1.2 IDENTIFICATION BY MIXED MELTING POINT

(electric melting point apparatus and mixed melting points)

Required

Two sets of semi-micro reagent bottles containing the following powdered substances (melting points in degrees Celsius shown in brackets), one set to be labelled and the others marked Y and Z only.

Benzene-1,2-dicarboxylic anhydride (131); carbamide (132)

Procedure

Identify Y and Z by carrying out a *single* mixed melting point determination (experiment 1.1, sequel 2). If available use the electric melting point apparatus.

1.3 MICRO SCALE DETERMINATION OF MELTING POINT

(hot stage microscope)

Required

Crystalline 2-hydroxybenzoic acid (m.p. 156 °C)

Procedure

1 Clamp the hot stage to the microscope and insert a 250 °C thermometer.

2 If the apparatus has not been used for some time, heat to about 140 °C to drive off any moisture.

3 Allow to cool and place a single crystal of 2-hydroxybenzoic acid on the glass support. Cover the crystal with a glass slip and place the baffle in position (fig 1.4).

4 Adjust the microscope mirror to illuminate the sample, and bring into sharp focus using the low power objective.

5 Turn on the heater and note the temperature range over which melting occurs.

1.4 PURIFICATION OF CAMPHOR BY SUBLIMATION

(simple funnel technique)

Required

Camphor; butanedioic acid

Procedure

1 Grind together a mixture of about nine parts of camphor to one part of butanedioic acid and record the melting point of the mixture.

2 Place 1 g of the mixture evenly over the bottom of a Petri dish and cover with a pierced filter paper and funnel (fig 1.6).

3 Gently warm the dish on a hot plate or over a small bunsen flame, being careful not to liquefy any of the mixture.

4 After an appreciable amount of camphor has collected in the funnel, allow the apparatus to cool. Scrape a few crystals of camphor on to a filter paper.

Sequel

1 Determine the melting point of the camphor (179 °C) and compare with that of the mixture. (Best results are obtained in this instance if the camphor is first melted in the melting-point tube and then allowed to solidify.)

2 Determine the melting point of the residue in the Petri dish (butanedioic acid, m.p. 185 °C).

1.5 SUBLIMATION OF ANTHRACENE-9,10-DIONE ON A SEMI-MICRO SCALE

(cold-finger assembly)

Required

Anthracene-9,10-dione; charcoal powder

Procedure

1 Grind together a mixture of ten parts anthracene-9,10-dione and one part of charcoal.

2 Introduce a little of the mixture into a small round-bottomed flask. as in fig 1.6.

3 Turn on the cooling water in the cold-finger and gently heat the flask. (There is no need to connect to the pump in this instance.)

4 Remove the cold-finger after a time, scrape off a little of the sublimate, and find its melting point using the electrical heater (anthracene-9,10-dione, m.p. 273 °C).

2 Boiling points

Although the forces between the molecules of a liquid are not powerful enough to produce a solid crystal lattice they are strong enough to prevent the substance behaving as a vapour. Even so the more energetic molecules are able to escape from the liquid surface (evaporation). Liquids which evaporate readily are said to be volatile (L. *volare,* to fly away). In an enclosed space equilibrium is reached between the molecules leaving the liquid phase and those re-entering it. The molecules escaping from the liquid in the form of a vapour exert a pressure in the same manner as the molecules of a gas. The pressure exerted by the vapour phase at equilibrium varies with the temperature and is known as the equilibrium vapour pressure. As the temperature rises there is a corresponding rise in the vapour pressure. When this just exceeds atmospheric pressure the liquid begins to boil at a constant temperature, known as the boiling point of the liquid, until evaporation is complete.

As with the melting point of a pure solid, the boiling point of a pure liquid is sharply defined. The presence of volatile or soluble impurities, however, causes fluctuations which are often of an erratic nature. The boiling point is also far more sensitive to alterations in atmospheric pressure than the melting point. For this reason pressure must be taken into consideration when using the boiling point as a criterion of purity or as a means of identification. In addition it is not uncommon to find mixtures of miscible liquids with a constant boiling point (azeotropic mixtures). Thus a mixture of 79.4% tetrachloromethane and 20.6% methanol boils steadily at 55.7 °C (see chapter 6).

The boiling points of members of an homologous series of organic compounds increase as their molecular weight rises. This is to be expected, as the larger molecules require more kinetic energy in order to escape from the liquid phase due to their greater mass and associated van der Waals' forces. For similar reasons associated polar compounds have a much higher boiling point than non-associated compounds of approximately the same molecular weight. This is strikingly illustrated in the case of the three compounds listed in table A which are of similar molecular weight.

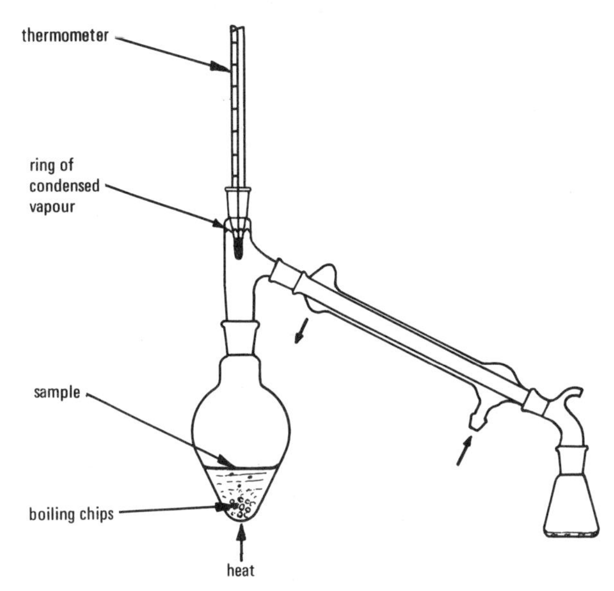

Fig 2.1 Semi-micro assembly for boiling point determination

Experimental details

The boiling point of an organic liquid may be conveniently determined using a semi-micro distillation assembly (fig 2.1). Apparatus with ground glass joints is strongly recommended, and the use of a bench scaffold instead of retort stands lessens the risk of strain due to poor alignment of the components. The distillation flask, preferably pear shaped, should be about one-third filled and should contain a few boiling chips to prevent superheating. After checking the range of the thermometer this should be positioned so that the bulb is opposite the condenser outlet. Heating is best carried out using a liquid bath or electric heating mantle and should be sufficient to maintain a ring of condensing vapour just above the level of the thermometer bulb. A satisfactory rate of distillation is indicated by the presence of a drop of condensed vapour on the end of the thermometer. The steady temperature at which the distillate passes over under these conditions may be taken as the boiling point of the liquid at atmospheric pressure.

Table A – Variation in boiling point of three compounds of similar molecular weight

Compound	Mol. wt.	Association	Polarity	b.p.(°C)
ethanol	46	associated	polar	78.8
methoxymethane	46	non-associated	slight polarity	–23.7
propane	44	non-associated	non-polar	–42.1

If the temperature of the vapour exceeds 140 °C it is advisable to drain the condenser of water and rely upon air cooling, otherwise the sharp temperature gradient at the edge of the water jacket might cause thermal stresses in the glass resulting in fracture.

When only very small quantities of a liquid sample are available and the determination of boiling point by distillation is impracticable, a semi-micro technique due to Siwoloboff can be employed. A melting-point tube (4 x 80 mm) sealed at one end is placed mouth downwards into a sample of the liquid under examination contained in an ignition tube. The ignition tube is attached to the stem of a thermometer of convenient range by means of a small rubber band (a slice from a piece of rubber tubing), the sample in the ignition tube being located alongside the bulb of the thermometer. The thermometer and attached tube are then gently warmed in a paraffin or silicone oil bath, care being taken that the rubber band is clear of the bath liquid (fig 2.2).

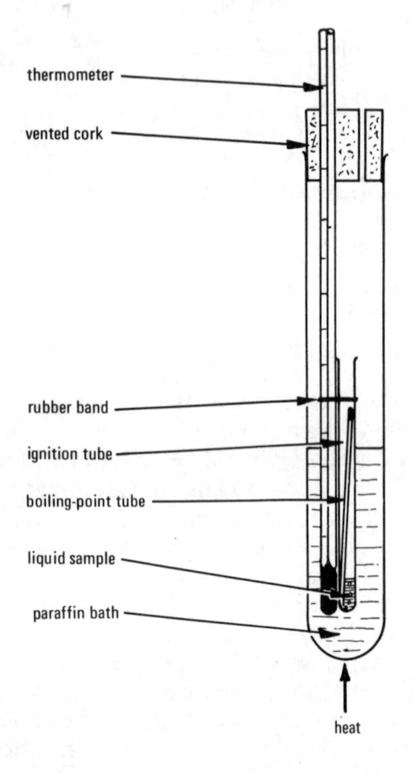

Fig 2.2

As the temperature rises, bubbles issue with increasing rapidity from the mouth of the melting-point tube until they form a continuous stream. At this point heating is stopped and the bath allowed to cool. The temperature at which the bubbles cease and the liquid starts to suck back into the tube is taken as the boiling point of the sample. Although not as accurate as the distillation method, the Siwoloboff technique gives a good enough approximation for most laboratory purposes.

The Emich technique for determining the boiling point of minute samples of liquid can be used as an alternative to the Siwoloboff method. A fine capillary of about 0.1 mm bore is formed by heating the centre portion of a 20 cm length of 1 mm capillary tubing and drawing out quickly. Two Emich tubes can be cut from this having a fine capillary tip about 1.5 cm in length.

By dipping the tip of the capillary into a sample of the liquid to be examined a drop is sucked into the end of the tube which is then sealed by briefly heating in a bunsen flame. In this way the drop of liquid encloses a small air bubble in the sealed capillary. The tube is then attached to a thermometer so that the drop is located alongside the bulb and heated in a liquid bath (fig 2.3).

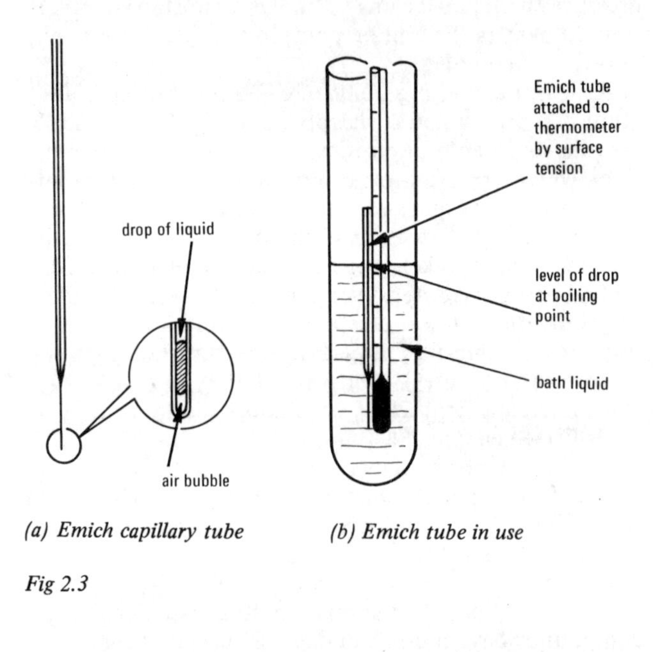

(a) Emich capillary tube *(b) Emich tube in use*

Fig 2.3

When the droplet starts to quiver the temperature of the bath is allowed to rise very slowly. The drop then moves up the capillary until it finally reaches the liquid level of the bath. At this point the bath temperature may be taken as the boiling point of the liquid.

Boiling point experiments

2.1 CONSTRUCTION OF A THERMOMETER CORRECTION GRAPH

(boiling point determination)

Required

Propanone (56); trichloromethane (61); tetrachloromethane (77); methylbenzene (110)

Broken porcelain or boiling chips

All liquids to be of analytical reagent quality (boiling points in degrees Celsius shown in brackets)

Procedure

1 Using the apparatus shown in fig 2.1, and a thermometer of range −10 to 110 °C, determine the boiling point of distilled water*.

2 Determine the boiling points of the four liquids provided*. Rinse the flask with a little of the liquid to be used before each determination.

3 Compare the observed boiling points with the values given above.

4 Construct a correction graph for the thermometer used as follows. Mark temperatures at 10°C intervals over the range 50–110 °C along a horizontal axis. Draw a vertical axis from −5 to +5 °C to represent the corrections at various temperatures. Plot the values obtained and join them with a smooth continuous line. The curve can now be used to give approximate thermometer corrections for temperatures within the limits of the scale.

2.2 VARIATION OF BOILING POINT WITH PRESSURE

(boiling point determination at reduced pressure)

Required

Ethyl ethanoate

Broken porcelain or boiling chips

Procedure

1 Determine the boiling point of ethyl ethanoate using the apparatus shown in fig 6.5, the manometer and pump being disconnected (page 23)*.

2 Connect the pump and manometer (see fig 6.6) and progressively reduce the pressure, noting down the temperature of the ethyl ethanoate vapour at convenient intervals.

3 At the end of the exercise, open the tap on the suction flask to equalize pressures and explain the effect on the contents of the distillation vessel.

* It is only necessary to adjust the boiling point reading if the barometric pressure is appreciably (more than 10 mm) above or below 760 mmHg, in which case reference must be made to boiling point tables (Kaye and Laby, *Tables of Physical and Chemical Constants*, Longman).

Sequel

Construct a vapour pressure diagram for ethyl ethanoate, plotting temperature in kelvins (horizontal axis) against pressure in mmHg (vertical axis).

2.3 MICRO SCALE DETERMINATION OF BOILING POINT

(Siwoloboff or Emich technique)

Required

Propanone (56); trichloromethane (61); tetrachloromethane (77); methylbenzene (110)

All liquids to be of analytical reagent quality (boiling points in degrees Celsius shown in brackets)

Procedure

1 Use the Siwoloboff (or Emich) technique (fig 2.2) to determine the boiling points of the four liquids provided.

2 Check your results with those obtained in experiment 2.1.

3 Filtration

Filtration is a means of separating the solid and liquid phases of a mixture. It is commonly used in organic chemistry to isolate crystallized substances from the mother liquor. Sometimes a sufficient degree of separation can be achieved by passing the liquid to be filtered through a plug of glass wool. One disadvantage of this method is that it is difficult to remove the residue completely from the glass fibres.

Paper is still the most commonly used of all filtration media. It can either be folded into the familiar cone or fluted form, or used as a flat disc according to the apparatus being used. For extraction purposes specially made paper 'thimbles' are used. On an industrial scale liquids are filtered by forcing through paper sheets or impregnated filter cloths.

Many different grades of filter paper are now available representing a wide range of porosity and thickness. This enables those papers to be selected which are most effective in separating particular types of suspension. Occasionally special filter discs made of black paper or woven glass fibre are used.

Free paper fibre ('floc') is also marketed in the form of compressed tablets or cellulose wool as a filter aid to accelerate the removal of very fine suspensions. When added to a suspension the floc forms a sludge of cellulose fibres which traps the solid particles and prevents them blocking the pores of the filtration medium. In general, however, the use of filter aids is more suited to inorganic separations.

Increasing use is being made of porous sintered glass plates for filtration, and more recently porous plastic sheet. These materials are chemically resistant, strong, and convenient, and although expensive compared with paper, they can be re-used almost indefinitely.

To speed up the filtration process both pressure and suction can be employed, although pressure filtration is almost entirely restricted to commercial processes. Most of the suction type filters used in the laboratory are variations on the Buchner funnel (fig 3.1).

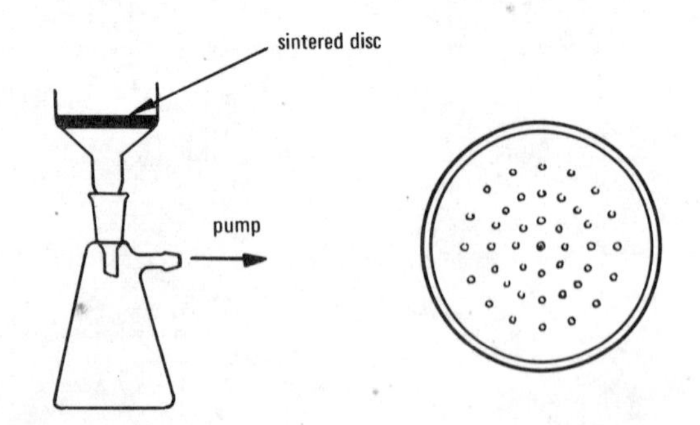

(a) All glass Buchner assembly *(b) Buchner-type sieve plate*

Fig 3.1

This consists of a cylindrical porcelain cup separated from the stem by a flat perforated disc which serves to support the filter paper. The stem of the funnel fits into the neck of a conical filtration flask with a short side arm to enable suction to be applied. Recent types of Buchner funnel are constructed of glass with either a slotted or sintered glass plate, and have the advantage of a ground joint stem which enables them to be mated directly to the suction flask. Sintered glass sieves are available in a number of different porosities ranging from coarse to very fine, a system of colour coded labels enabling the correct grade to be selected at a glance. For routine laboratory work a Buchner funnel of about 42 mm diameter is convenient.

Small scale methods

For small scale filtration a specially modified Buchner funnel termed a Hirsch funnel is commonly used. This has sloping sides and a small perforated sieve plate just above the origin of the hollow stem. The smallest Hirsch funnels have plates which are only a few millimetres in diameter. Glass Hirsch type funnels are available fitted with sintered glass discs. Suction is applied by mounting the funnel in a filter tube provided with a side arm. The filtrate is conveniently collected in a centrifuge tube resting on a pad of glass wool in the bottom of the filter tube (fig 3.2a).

A useful modification of the simple suction tube is the Irvine filter cylinder (fig 3.2a). This enables the filtrate to be drawn off without removing the funnel or interrupting the suction.

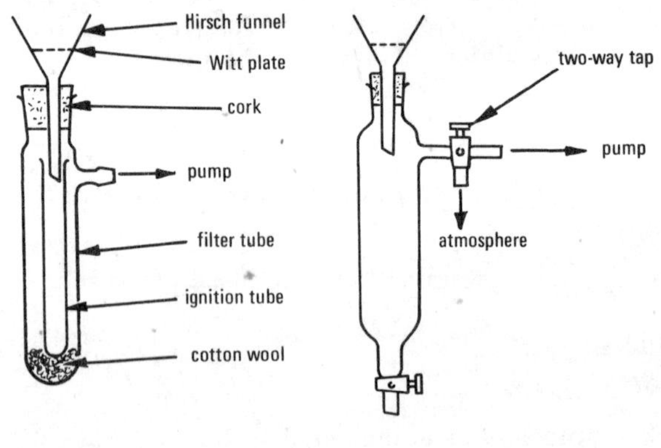

(i) Hirsch filtration assembly *(ii) Irvine cylinder*

Fig 3.2 (a)

Another convenient device for the filtration of small volumes of liquid is the Willstätter nail (fig 3.3a). This is a finely tapered glass rod which has been flattened and slightly corrugated at the broad end by pressing on a coarse file after heat softening. The 'nail' fits into the stem of a small glass funnel which usually has a ground joint to enable it to be fitted into the neck of a filter tube. Before use, the head of the nail is covered by a small disc cut from a sheet of filter paper with a clean, sharp cork borer.

A Pregl filter (fig 3.2b) can be made by fitting the narrow end of an adsorption tube into the mouth of a filter tube using a rubber stopper. The bulb of the adsorption tube can then be packed with glass or cotton-wool and filtration carried out by suction.

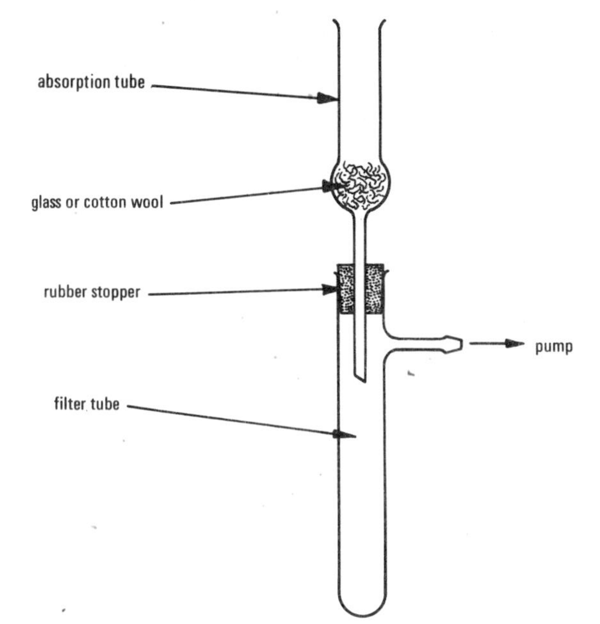

Fig 3.2 (b) Pregl filter

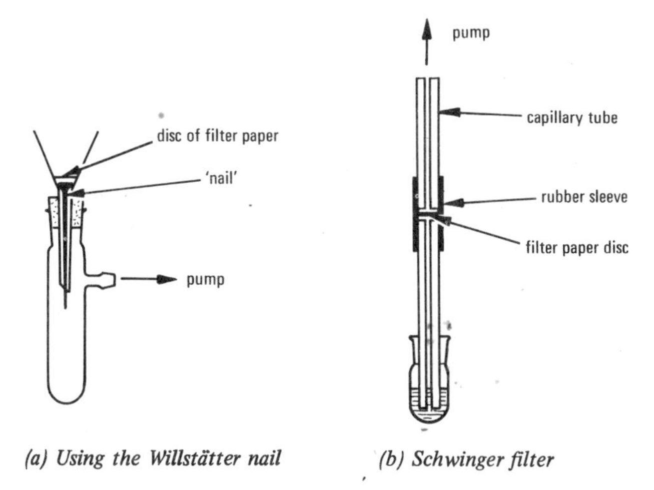

(a) Using the Willstätter nail *(b) Schwinger filter*

Fig 3.3

For micro scale filtration a Schwinger filter is used (fig 3.3b). This consists of two short lengths of capillary tubing separated by a small disc of filter paper and held together by a sleeve of rubber tubing. The lower end of the capillary tubing is placed in the liquid to be filtered and upward suction applied. The residue can be washed and dried *in situ* by drawing distilled water or other suitable solvent over it followed by hot air.

Experimental details

When using discs of filter paper these should be wetted with a little of the solution to be filtered before applying suction, to ensure an intimate contact with the filter. Although it is not necessary to use a paper disc with a sintered glass filter, it is often convenient to do so as this facilitates removal of the solid residue when filtration is complete. Soft-walled tubing should never be used to connect the suction vessel with the pump since it will collapse as the internal pressure is reduced. Stiff plastics tubing is ideal for this purpose, thick-walled rubber tubing being too heavy for small-scale work. It is wise to hold the filtration flask in a clamp while filtration is in progress to avoid it being overset by the weight of the heavy suction tubing.

Suction should not be applied too fiercely as this often blocks the pores of the filter paper with fine particles and may even rupture it. Care must also be taken not to turn off the suction suddenly when using a water pump or else 'suck-back' will occur. After all the solution has been filtered, the residue may be washed to remove traces of mother liquor using a little cold solvent. The solid is then dried out by gentle suction for a few minutes after the wash liquor has passed through. If there is a substantial amount of solid residue this should be gently tamped down with an inverted glass stopper during the final drying stage to prevent cracks appearing which would cause loss of suction. Alternatively a piece of rubber balloon can be stretched loosely over the mouth of the funnel and held in place with an elastic band. This will be sucked down onto the surface of the residue and prevent cracking taking place.

To remove the residue from a Hirsch or Buchner funnel, the latter is inverted over a sheet of thick filter paper and gently tapped on the bench. The filter paper invariably comes away with the residue but can easily be peeled off. In the case of the Willstätter filter the residue is readily removed by pressing on the projecting end of the nail.

It is often necessary when purifying a substance by crystallization to filter the solution hot, thus removing insoluble impurities. To avoid premature crystallization it is essential to heat the funnel and carry out the filtration as rapidly as possible. In addition the funnel stem must be short.

When using an ordinary glass funnel it is sufficient to use a fluted filter paper (fig 3.4) and to pre-heat it by

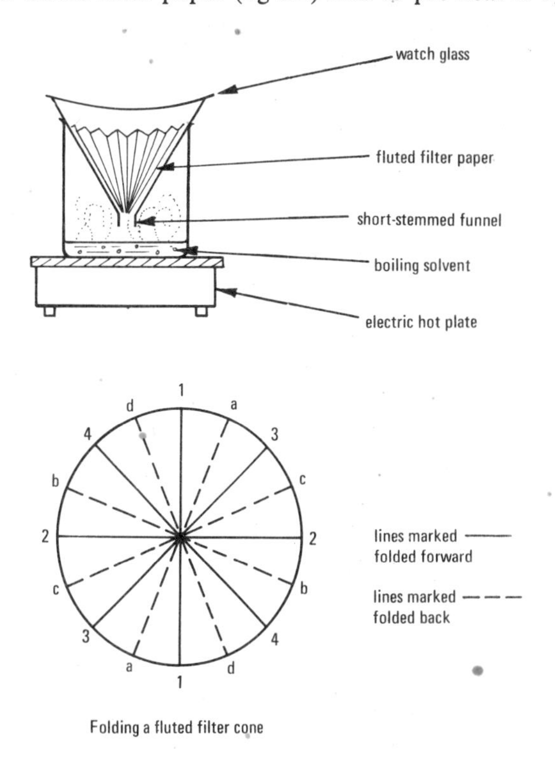

Fig 3.4

pouring through a little hot solvent immediately before use. A more effective method is to rest the funnel in the mouth of a small beaker containing a little boiling solvent (fig 3.4). In this way the funnel and filter paper are bathed in hot vapour during the filtration. Excess solvent is easily removed by evaporating the filtrate down to the required volume before crystallizing.

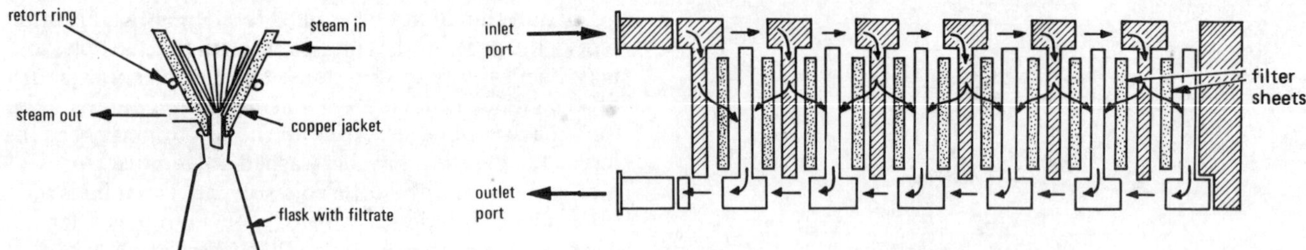

(a) Steam funnel heater

Fig 3.6 Exploded diagram of a filter press

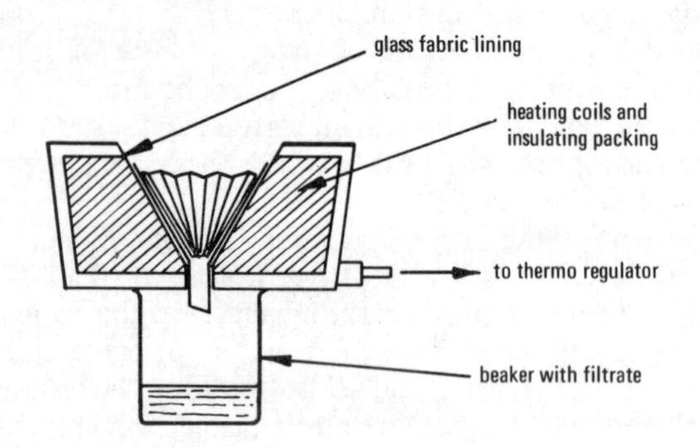

(b) Electrical funnel heater

Fig 3.5

For larger quantities of liquid, or solutions which readily crystallize on cooling, special funnel heaters are necessary (fig 3.5). Older types of heater use hot water or steam circulating in a copper jacket or coil around the funnel. Electric funnel heating mantles are more expensive but give a wider range of temperatures and can be easily controlled by an isolated thermoregulator. They are also safer to use when handling flammable solvents.

Industrial filtration

Filtering processes play an important part in industry. Special techniques and plant are required to deal rapidly and efficiently with the very large volumes of liquid which are often involved.

Filter presses have been used since the earliest days of the chemical industry. The modern plate and frame filter press consists of a number of recessed metal plates known as 'leaves'. Sheets of finely woven filter cloths are placed between the leaves which are then tightly pressed together in a frame by means of a hydraulic press or screw. The liquid to be filtered is forced under pressure along the channels in the plates. The filtrate passes through the filter cloth leaving a cake of solid residue and is run off from the bottom of the press (fig 3.6).

It is common practice to coat the filter cloth with a layer of diatomaceous earth or other filtration aid before use and to add similar materials to the liquid to be filtered. This is to coagulate any slimy material which would block the filter, and absorb colloidal particles which would otherwise pass through the filter cloth. The shape and size of the particles of the filter aid are of great importance in producing optimum results. Filter presses are widely used in the manufacture of such products as dyestuffs, varnishes, fats, and pharmaceuticals, and in the refining of sugar and the clarification of beers and wines.

A disadvantage of the filter press is that the process is discontinuous. The press has to be opened at intervals to remove the solid residue (filter cake) from the filter cloths. This difficulty is overcome by the use of continuous drum filters. The drum revolves horizontally on its axis, and has a perforated periphery which is usually covered with a filter cloth. The lower part dips into the liquid to be filtered and this is sucked into the drum, leaving a layer of solid on the exterior. As the drum revolves the solid is washed and then continuously removed by a scraper while the filtrate is run off from the inside (fig 3.7). A thin

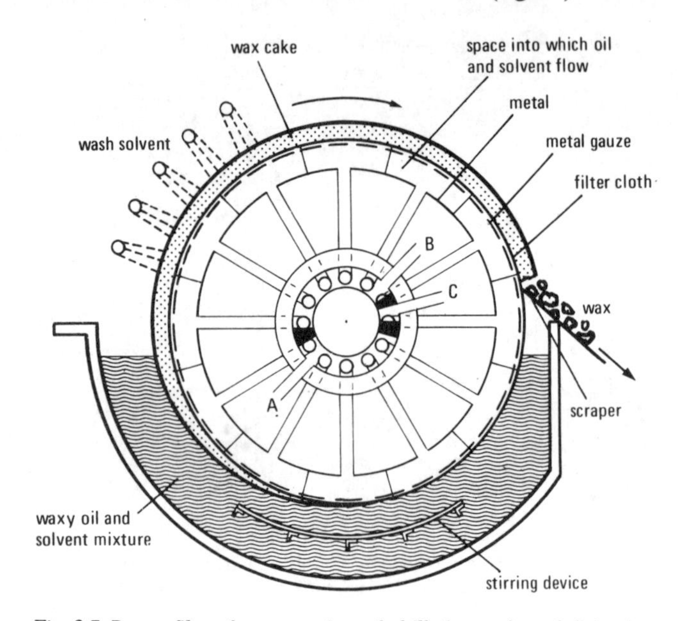

Fig 3.7 Drum filter for separation of chilled wax from lubricating oil

layer of material is left on the drum surface as this provides a suitable bed for a fresh layer of solid. This type of filter has proved very effective in such processes as the removal of paraffin wax from chilled lubricating oil, and in the filtration of antibiotic mould cultures and yeast suspensions.

Another type of continuous filter consists of a number of slowly revolving discs which dip into the liquid to be filtered and operate in a similar way to the drum filter.

For the removal of very fine precipitates, 'streamline' filters are used. These consist of wads of filter papers pressed tightly together, the liquid being forced through the edges of the sheets. For small scale filtration, such as that required for removing contaminants from engine oil, 'by-pass' filters like the Metafilter are used. In these the oil is forced from a central perforated cylinder through a number of sheets of filter paper impregnated with diatomaceous earth. This type of filter is ingeniously designed to give a filtration area of about 1.3 m^2, although the discs are only about 10 cm in diameter.

Filtration experiments

3.1 PURIFICATION OF CONTAMINATED ASPIRIN

(hot filtration and Hirsch filter)

Required

Dilute ethanoic acid; contaminated aspirin tablets*

Procedure

1 Fit a fluted filter paper into a short-stemmed funnel and rest the funnel in a small beaker.

2 One-quarter fill a boiling tube with dilute ethanoic acid, add two tablets of contaminated aspirin, and heat until dissolved.

3 Pour the hot aspirin solution quickly into the fluted filter.

4 Allow the filtrate to cool and crystallize.

5 Filter off the crystallized aspirin using a Hirsch filter (fig 3.2a).

6 Wash the crystals with a little cold water and suck dry.

Sequel

1 Select one or two crystals and examine them under low power magnification.

2 Dry the remaining crystals on filter paper and record the yield and melting point (aspirin 136–137 °C).

3.2 PREPARATION OF TRIIODOMETHANE ON A SEMI-MICRO SCALE

(Willstätter nail filter)

Required

Propanone; solution of sodium hydroxide (10%); solution of potassium iodide (10%); industrial methylated spirit (ethanol); commercial sodium chlorate(I) [sodium(I) oxochlorate(I)] solution

Procedure

1 Mix together in a 100 cm^3 conical flask 1 cm^3 of propanone, 5 cm^3 of 10% sodium hydroxide solution and 10 cm^3 of 10% potassium iodide solution.

2 Add 10 cm^3 commercial sodium chlorate(I) solution and shake well.

3 Allow to stand for a few moments and then decant off as much liquid as possible.

4 Shake with a little distilled water and then filter off the yellow suspension of triiodomethane using a Willstätter nail filter (fig 3.3).

5 Dissolve the triiodomethane in a minimum of boiling methylated spirit (use a hot plate or water bath), and cool.

6 Filter off the crystals using the Willstätter nail, wash with a little cold methylated spirit, and suck dry.

Sequel

1 Select one or two crystals and examine them under low power magnification.

2 Dry the remaining crystals on filter paper and record the yield and melting point (triiodomethane m.p. 119-120 °C.

3 Note the characteristic odour of triiodomethane, formerly used as an antiseptic.

3.3 ISOLATION OF PEROXIDASE FROM TURNIPS

(Buchner filter)

Required

One small turnip

Ammonium sulphate(VI); phenylamine;
2-methylphenylamine; 4-methylphenylamine;
glacial ethanoic acid; hydrogen peroxide (20 vol)

Procedure

1 Top and tail a small turnip, and pulp by rubbing on the side of a kitchen grater.

2 Place the turnip pulp in a large mortar, cover well with water and grind for a few minutes with a pestle.

3 Decant off the liquid into a Buchner funnel (fig 3.1) and apply suction.

4 Transfer the pulp to the funnel and press down with an inverted glass stopper to extract as much juice as possible.

5 Pour the filtrate into a beaker and stir in 7 g of ammonium sulphate(VI) for each 10 cm^3 of extract. Allow to stand for some minutes.

6 Filter off the precipitate of crude peroxidase using a Buchner funnel, wash with a little cold water and suck dry.

7 Grind the peroxidase with a little water in a mortar and filter through a fluted paper (fig 3.4) into a boiling tube.

Sequel

1 Add two drops of phenylamine, 2-methylphenylamine and 4-methylphenylamine respectively to three test tubes. Dissolve the samples in an equal volume of ethanoic acid and then add five drops of the peroxidase extract and two drops of the 20 vol hydrogen peroxide. Record your observations.

2 Repeat the experiment with peroxidase solution which has been boiled for a few minutes and comment on the result.

*Use non-soluble aspirin and contaminate by spotting each tablet with a drop of ink and then rubbing on a dusty surface.

3.4 HYDROLYSIS OF ASPIRIN

(Hirsch filter)

Required

Aspirin; sodium hydroxide solution (10%, 2.5 M)

Procedure

1 Add 20 cm^3 of 10% sodium hydroxide solution to 1 g of aspirin in a 50 cm^3 conical flask.

3 Drop in a few porcelain chips, boil gently for 15 minutes and then allow to cool.

3 Acidify the cold solution with dilute sulphuric(VI) acid and filter off the precipitate of 2-hydroxybenzoic acid using a Hirsch filter (fig 3.2a).

Sequel

1 Select one or two crystals and examine them under low power magnification.

2 Dry the remaining crystals on filter paper and record the yield and melting point (2-hydroxybenzoic acid m.p. 150 °C).

3 Shake a few of the 2-hydroxybenzoic acid crystals with a little water in a test tube and add a few drops of iron(III) chloride solution. Repeat with a little aspirin solution in water and compare the results.

4 Centrifugation

Suspensions of insoluble particles or droplets often separate out under the influence of gravity, the time taken for this to happen depending upon the weight and size of the particles. In the case of very fine suspensions however, the force of gravity alone is not powerful enough to overcome the opposing forces due to diffusion, viscosity, and with emulsions, the mutual repulsion of the electrically charged droplets. Consequently centrifugal force is used to simulate a greatly increased gravitational field by spinning the sample to be separated at very high speed. Thus, using a rotor arm of 10 cm radius revolving at 1200 rev/min a force equivalent to 164*g* (164 times the earth's gravitational field) is generated. Under these conditions, even the finest suspensions are quickly and effectively separated. This technique, known as centrifugation, was first used in the extraction of tung oil over a thousand years ago.

Besides its value as a method of separation, the centrifuge can give information concerning the physical properties of the particles themselves. In this way the sedimentation time of large molecules such as proteins during high speed centrifugation has provided an elegant method for the determination of their molecular weights.

Typically the centrifuge consists of a rotor which is spun at high speed by an electric motor or turbine. The rotor carries the samples to be separated in tiny buckets, and these are carefully balanced to avoid vibration and wear on the bearings. A protective metal cover surrounds the device to avoid accidents during operation.

The centrifugal force developed is proportional to the square of the radial velocity and to the radius of the rotor. At very high speeds the forces developed may be great enough to cause the rotor to disintegrate. This imposes a physical limit on the speed of the centrifuge, but a limit that is continually being extended by careful design. The early pioneer work on centrifuge design was carried out by Thé Svedberg, Professor of Physical Chemistry at the Swedish University of Uppsala. In 1923, he designed an electrically driven 'optical' centrifuge which enabled the process of sedimentation to be observed and photographed through a window in the rotor. Difficulties were caused by thermal stirring due to convection currents set up by the friction heated rotor, and swirling caused by the deceleration forces produced when the centrifuge was stopped. These were partly overcome by the use of narrow sample tubes and a turbine driven rotor which was spun in a low pressure atmosphere of hydrogen. The use of hydrogen not only cut down frictional heating, but provided a coolant for the rotor.

In the laboratory, centrifuges are widely used for the concentration or separation of suspensions, and the 'breaking' of emulsions. The sample to be centrifuged is placed in a narrow glass centrifuge tube with a conical extremity, which is carried in a swinging metal bucket hanging from the rotor arm. Normally the rotor carries from two to eight of these buckets which swing up into the horizontal position during centrifugation. This is usually carried out at about 2000 rev/min. In recent models the rotor casing is designed to swing with the rotor in order to cut down air friction. For higher speeds (up to 17000 rev/min) the rotor is in the form of a solid bowl which is drilled to receive plastic centrifuge tubes. In addition the centrifuge cabinet is sometimes air evacuated and cooled before use.

Laboratory centrifuge

Alternative design of rotor head

Experimental details

Care must be taken not to overfill sample tubes which should not be more than three-quarters full. Drops of liquid should not be allowed to contaminate the buckets or the rotor, and the centrifuge should be evenly loaded using opposing pairs of buckets. If only one bucket is being used then it must be counterbalanced by a tube containing an equivalent amount of water. The lid of the centrifuge should be closed before spinning and not reopened until the rotor has stopped. On no account should the lid be opened during operation, or attempts made to decelerate the rotor quickly by digital pressure on the centre pivot. If any unusual noise or vibration is experienced during the operation of the centrifuge it should be switched off immediately and the cause investigated.

Antibiotics production: close-up of a centrifugal extractor

Many centrifuges for laboratory use are fitted with a press switch so that it is impossible to leave the rotor spinning in the absence of the operator.

The length of time the sample should be spun can only be judged by experience. The sediment should be packed solidly enough to allow supernatant liquor to be removed completely with a small teat pipette. Firm packing of the sediment also prevents remixing during deceleration of the rotor.

Industrial centrifuges

The simplest pattern of industrial centrifuge operates in a manner similar to the domestic spin dryer. A perforated rotating drum is used to throw off liquid from the solid phase of a mixture. If the solid phase is finely divided it is necessary to line the drum with a filter cloth. Relatively slow speeds of about 1000 rev/min are used to generate centrifugal forces equivalent to about 200 *g*. Sugar crystals are separated from mother liquor in this manner.

In common with most other industrial processes it is more economical to use a centrifuge continuously than to work on a batch system. For this reason hollow cylindrical centrifuges have been developed which can be operated continuously. The liquid to be separated is fed into the lower end of the cylinder which is spun on its vertical axis. The liquid climbs the wall of the centrifuge in a thin layer and is continuously removed from an overflow port, while the denser sediment is pressed tightly against the cylinder wall. This type of centrifuge has been used to recover the pentyl ethanoate solvent used in the extraction of penicillin from culture solutions.

A more efficient design of continuous centrifuge contains a number of conical plates which subdivide the liquid into layers. This speeds up the process of separation and minimizes the risk of re-mixing. Cream separators are usually of this type.

The ultracentrifuge

Centrifuges with speeds in excess of 20000 rev/min are called ultracentrifuges, a term originally used by Svedberg. Enormous gravitational forces can be developed in these machines which are designed to give information about the physical nature of minute particles by sedimentation techniques.

Already, valuable details have been obtained about the structure of such minute particles as protein molecules and viruses. In 1926, Svedberg designed the first high speed centrifuge which was rotated by means of an oil driven turbine. Using a 5 cm rotor arm revolving at 45000 rev/min a centrifugal force in excess of 100000 *g* was obtained. By 1934, Svedberg and his team were experimenting with centrifuges developing a gravitational force approaching a million *g*, but the immense strain on the rotor caused frequent disintegrations.

Meanwhile two French scientists, Henriot and Huguenard, were experimenting with solid bowl-type rotors driven by an air turbine and supported on an air cushion while spinning. Using this principle, Pickels and Beams, working at the University of Virginia, designed a novel type of rotor which was suspended from the turbine by a length of piano wire and spun in an evacuated container.

Beams has also designed a coasting centrifuge. This has an 18 cm rotor operating in a vacuum. When the speed reaches 60 000 rev/min the turbine drive drops off the rotor which is suspended in mid-air by a powerful magnetic field. The frictional losses of such a device are extremely small and the rotor 'free-wheels' under this condition for more than a year before coming to rest.

The rate of sedimentation of particles in the ultracentrifuge is photographed by white or ultraviolet light using a novel technique. This makes use of the refracted light which appears at the boundary between layers of different density.

An interesting application of ultracentrifugation on an industrial scale has been in the field of uranium enrichment. Proposals have been made to build a joint Dutch, British, and German installation with a potential capacity of three kilotonnes per year. Workers at the Tokyo Institute of Technology have also used a centrifuge spinning at 25000 rev/min to enrich natural uranium.

Centrifugation experiments

4.1 EMULSION BREAKING BY CENTRIFUGATION

(laboratory centrifuge)

Required

Turpentine oil; cedarwood oil; fresh milk; salad cream; hand lotion

Procedure

1 Vigorously shake a little turpentine with an equal volume of distilled water in a stoppered conical flask and pour a little of the resulting emulsion into a test tube.

2 Repeat the operation with cedarwood oil and water.

3 Three-quarters fill one centrifuge tube with the turpentine oil emulsion and another with the cedarwood oil emulsion, and spin in opposing buckets for 2–3 minutes. Examine and compare with the untreated samples.

4 Repeat the experiment with a sample of fresh milk from a bottle which has been well shaken, and salad cream (an emulsion of olive oil and vinegar stabilized with egg yolk).

5 Spin a cosmetic emulsion such as hand lotion that contains a powerful emulsifying agent.

Sequel

Tabulate your results and comment on the limitations of using centrifugation for breaking emulsions.

4.2 USE OF CENTRIFUGATION IN SEDIMENTATION

(laboratory centrifuge)

Required

Cotton wool; methylbenzene; detergent solution; metal polish; disperse dye such as Dispersol (ICI); dilute sulphuric(VI) acid; barium(II) chloride; hydrochloric acid; disodium(I) thiosulphate(VI); Aquadag

Procedure

1 Ignite a *wisp* of cotton wool dampened with a *few* drops of methylbenzene. Hold a test tube filled with water in the flame to obtain a coating of fine soot.

2 Wash off some of the soot into a boiling tube containing water with a few added drops of detergent solution.

3 Three-quarters fill two centrifuge tubes with the sooty water and spin in opposing buckets. Stand the tube with the remainder of the liquid in a test tube rack.

4 After spinning for 2–3 minutes compare the centrifuged samples with the untreated liquid.

3 Repeat the experiment using a well shaken sample of metal polish and an aqueous suspension of a disperse dye.

6 Three further dispersions of colloidal dimensions can be prepared as follows and used for demonstrating the effect of centrifugation:

(i) a precipitate of barium(II) sulphate(VI) produced by adding dilute sulphuric(VI) acid to aqueous barium(II) chloride;

(ii) a colloidal suspension of sulphur produced by adding a few drops of concentrated hydrochloric acid to a solution of disodium(I) thiosulphate(VI);

(iii) a colloidal suspension of graphite produced by stirring a little Aquadag into water.

4.3 SEPARATION OF METHYL ORANGE AND METHYLENE BLUE BY MICRO-CHROMATOGRAPHY

(laboratory centrifuge and micro-adsorption column)

Required

Aluminium(III) oxide (chromatographic grade); methylene blue; methyl orange; ethanol (industrial methylated spirit); cotton wool

Procedure

1 Prepare a micro-column as illustrated in fig 8.4 (page 38).

2 Push a small wisp of cotton wool down the tube to act as a plug, and put the narrow end of the column through the plastic cap of a snap top bottle.

3 Tap small amounts of aluminium(III) oxide down the tube until it is half full.

4 Dissolve a little methylene blue and methyl orange in ethanol and run one drop down the tube onto the surface of the column.

5 Fill up the tube with ethanol, and place the bottle and tube into a centrifuge bucket containing a slice of rubber at the bottom, as shown in the diagram.

6 Centrifuge for a minute and then examine the chromatogram produced. Spin again if necessary.

Sequel

1 See if you can remove the methylene blue completely by elution with the ethanol, and then remove the methyl orange by elution with water.

2 Pour the eluted fractions into two large watch glasses and allow the solvent to evaporate so that samples of the original dyes can be recovered.

5 Crystallization

The purification of solids by crystallization is an elegant and effective technique of great antiquity. It is still widely used by the chemist, especially in the organic field. The method is mainly dependent upon the simple fact that most solids are more readily soluble in a hot solvent than a cold one. In addition any small amounts of soluble impurities remain dissolved in the mother liquor as pure solute crystallizes out from a saturated solution. This selective crystallization is aided by the fact that the orientation of molecules into a crystalline lattice is highly specific, and will only rarely include molecules of other substances.

Typically the impure material is completely dissolved in a minimum amount of boiling solvent. This is filtered to remove any insoluble solid impurities, and the solution allowed to cool. Crystals of the pure substance separate out, and are washed and dried.

Occasionally the presence of traces of impurity causes the crude material to be discoloured. This can usually be remedied by shaking the hot solution with a little animal charcoal before filtration. Impurities present in a finely divided state can be removed in a similar fashion.

Crystallization can be carried out by simply preparing a solution of the solid, and allowing the solvent to evaporate spontaneously. This is much less effective, however, and usually leads to the formation of a surface crust which includes the impurities.

Choice of solvent

The successful application of crystallization techniques depends very largely upon the correct choice of solvent. The relationship between specific solvents and their behaviour is very complex. Selection of a solvent is therefore usually carried out by experiment. Nevertheless there are a few generalizations regarding solvent behaviour which are helpful in making a choice.

The old dictum that 'like dissolves like' is a useful guide. Many organic compounds do not ionize in water. It is only possible for them to dissolve in water if their molecules contain functional groups such as carbonyl (>C=O) and hydroxyl (—OH) which are able to form hydrogen bonds with the water molecules. Thus hydrocarbons such as paraffin wax are insoluble in water but amides such as ethanamide are water soluble.

Organic substances which do not possess hydrogen bonding properties usually dissolve in solvents of a similar nature, e.g. ethoxyethane and benzene. In this instance solution is effected by simple molecular mixing of solute and solvent. Solvents which possess a hydroxyl group are often associated and have solvent properties intermediate to water and the hydrocarbon type solvents. Methanol and ethanoic acid fall into this category (table B).

It often happens that a solid is readily soluble in one solvent and almost insoluble in another, so that neither solvent is suitable for crystallization. In this case a mixture of the two solvents can often be used most effectively. A solution of the solid is first prepared by heating with the more powerful solvent. While the solution is still hot a little of the second solvent is added drop by drop until the solution becomes slightly turbid. The turbidity is then dispelled by a few drops of the first solvent, and the solution allowed to cool as before, when crystallization readily occurs. Some examples of mixed solvents are ethanol/water, methanol/water, ethanoic acid/water, ethanol/benzene*, and benzene/petroleum ether*.

Ideally a solvent should be safe to use and economical, readily dissolving the solute when hot but only sparingly in the cold.

*Highly flammable. Whenever benzene is used extra care should be taken because of danger from skin absorption and inhalation.

Table B – Some common solvents

Miscible with water	b.p.(°C)	Immiscible with water	b.p.(°C)
industrial methylated spirit †	77-82	benzene††	81
methanol†	65	petroleum ether†	40-120
propanone†	56	trichloromethane	61
ethanoic acid (glacial)	118	tetrachloromethane	77

†Highly flammable.

††Highly flammable. Danger from skin absorption and inhalation.

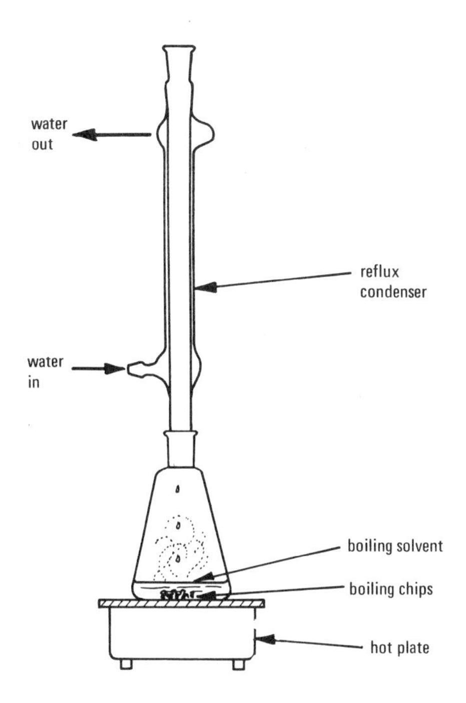

Fig 5.1

Experimental details

Solution

The finely powdered solid is placed in a suitably sized conical flask with a few boiling chips (aluminium(III) oxide anti-bumping granules are convenient). Slightly less than the estimated volume of solvent required is added and solution effected by heating on a water bath or hot plate. If the solvent is expensive, toxic, or flammable, a reflux condenser should be fitted to the flask (fig 5.1). The gently boiling mixture is swirled around and small additional quantities of solvent added until a clear solution is obtained. The presence of small amounts of insoluble impurity can usually be easily recognized. Decolorization is carried out at this stage by adding a little animal charcoal. Great care must be taken not to add a large excess of charcoal as this leads to substantial losses of liquid and solute by adsorption.

Filtration (see page 9)

The solution must be filtered hot, and to avoid premature crystallization the funnel should be short-stemmed and warmed before use by pouring through a little boiling solvent. The use of a fluted filter paper also speeds up filtration. It is convenient to use a funnel heater, but with the electrically heated type care must be taken that the temperature does not exceed the b.p. of the solvent. For larger volumes of liquid which do not crystallize rapidly on cooling, a small Buchner or Hirsch funnel can be used.

The filtrate is best contained in a small conical flask and covered with a watch glass or inverted beaker. This prevents evaporation during cooling and also keeps out dust.

Crystallization

The size of crystals depends upon their rate of growth. Slow cooling produces larger crystals, while rapid cooling with stirring produces much finer crystals. Since most organic compounds tend to form small crystals, slow cooling is advisable to facilitate their recovery. When crystallization is complete, the crystals are removed from the mother liquor by suction filtration, and then rinsed with a little fresh chilled solvent to remove any adhering solution. With small samples a Willstätter nail can be used.

Occasionally, on cooling the hot solution an oily syrup is produced which resists crystallization. If this happens the oil should be redissolved by heating, and the solution allowed to cool while the inside surface of the container is scratched with a glass rod.

Drying

After recovery the crystals must be dried. As much liquid as possible is removed by filtration. The funnel is then inverted on two or three sheets of filter paper, and the cake of crystals dislodged by a sharp tap. Most of the remaining liquid can now be removed by gently pressing the crystals between clean sheets of thick filter paper, the soiled sheets being replaced as necessary.

The last traces of water may be removed by transferring the crystals to a watch glass and placing in a desiccator. This is charged with a powerful dehydrating agent such as silica gel or anhydrous calcium(II) chloride. 'Self-indicating' silica gel granules are available which change colour from blue to pink when they require regenerating. The spent granules can be dehydrated by gentle heating in an oven and re-used.

A useful combination for drying a wide range of materials is to have concentrated sulphuric(VI) acid in the base of the desiccator and sodium(I) hydroxide flake in the collar. The presence of a few freshly prepared shavings of paraffin wax is effective in removing traces of hydrocarbon solvents such as benzene or petroleum.

Drying at atmospheric pressure is a slow process and the removal of traces of liquid occurs much more rapidly at reduced pressures. For this reason a vacuum desiccator is often used which can be connected to a suction pump (fig 5.2).

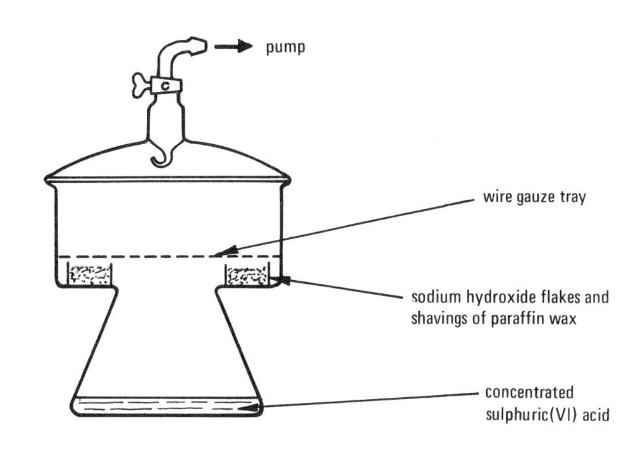

Fig 5.2 Vacuum desiccator

A steam oven, or better still a thermostatically controlled electric oven, can be used to dry crystal samples quickly. Care must be taken that the temperature used is well below that of the melting point of the solid. Vacuum ovens are sometimes used to drive off tenaciously held liquid.

A convenient form of apparatus for effectively drying

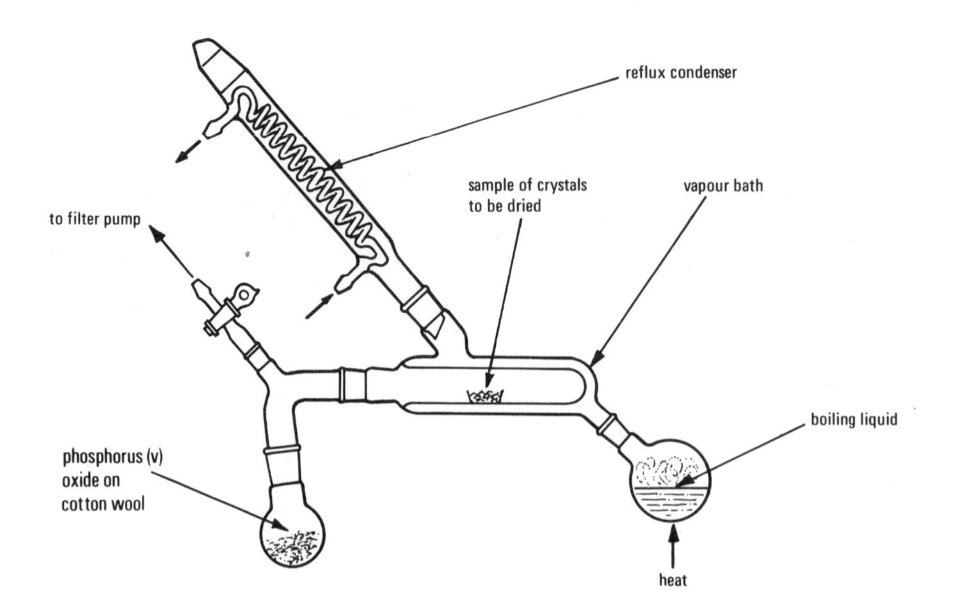

Fig 5.3 A drying 'pistol'

crystals which makes use of both heat and vacuum is the Abderhalden drying 'pistol' (fig 5.3). In this the crystals are placed in an inner tube which is surrounded by hot vapour from a suitable liquid boiling under reflux. The inner tube is connected to a drying unit containing phosphorus(V) oxide and this is in turn connected to a filter pump. An electrically heated version of the drying pistol is also in common use.

Crystallization experiments

5.1 PURIFICATION OF CRUDE N-PHENYLETHANAMIDE

(recrystallization and decolorization)

Required

Crude N-phenylethanamide (technical grade); animal charcoal; demerara sugar; vinegar; Esso Blue domestic paraffin

Procedure

1 Dissolve 2 g of crude N-phenylethanamide in a minimum amount of boiling water in a 100 cm^3 conical flask. The presence of a dark oil indicates undissolved N-phenylethanamide.

2 Add an additional 5 cm^3 of water at boiling point together with a little animal charcoal. Boil gently for five minutes and filter through a heated funnel containing a fluted filter paper (fig 3.4).

3 When the filtrate is cold and crystallization complete, collect the crystals by suction filtration through a Buchner funnel.

4 Wash the crystals twice with cold water and dry between filter papers.

Sequel

1 Record the yield and the melting point (N-phenylethanamide m.p. 114 °C).

2 Activate 3 or 4 g of animal charcoal by heating in a tin dish. Add 1 g of activated charcoal to 50 cm^3 of demerara sugar solution and reflux (fig 5.1) for about five minutes. Filter by suction using a Buchner funnel. Repeat using vinegar and Esso Blue paraffin. Comment on your results.

5.2 RECRYSTALLIZATION OF IMPURE 1,3-DINITROBENZENE

(recrystallization using a volatile solvent)

Required

Crude 1,3-dinitrobenzene (technical grade); ethanol (industrial methylated spirit)

Procedure

1 Add 8 cm^3 of ethanol to 0.8 g of technical grade 1,3-dinitrobenzene in a small conical flask.

2 Reflux until dissolved (fig 5.1) adding further ethanol via the condenser tube if required.

3 Filter the solution hot through a fluted filter paper (fig 3.4) and allow to cool and crystallize.

4 Filter off the crystals using a Hirsch funnel and wash with a few drops of cold ethanol.

Sequel

1 Examine one or two crystals under low power magnification. Do these differ from the crude salt?

2 Dry the remainder of the crystals between filter papers and check the yield and melting point (1,3-dinitrobenzene m.p. 89 °C).

5.3 PREPARATION OF OSAZONES FROM GLUCOSE AND LACTOSE

(recrystallization on a semi-micro scale)

Required

Glucose; lactose; phenylhydrazine; ethanoic acid; ethanol (industrial methylated spirit)

Procedure

1 Dissolve 0.5 g of glucose in 5 cm^3 of water in a test tube. In a second tube dissolve 1 g of phenylhydrazine in 1 cm^3 of ethanoic acid and dilute with water to about 10 cm^3.

2 Mix the two solutions and warm the tube containing the mixture in a beaker of boiling water.

3 After heating for 15 minutes allow the tube to cool and filter the contents using a Willstätter nail.

4 Wash the crystals with water followed by a few drops of ethanol.

5 Recrystallize by refluxing the glucosazone with a minimum of ethanol (fig 5.1) and then allowing to cool.

Sequel

1 Examine the crystal structure of glucosazone under low power magnification and sketch.

2 Repeat the experiment using lactose. For recrystallization boil the lactosazone with a little water and add ethanol drop by drop until the solution clears, then cool slowly.

3 Examine the crystal structure and sketch as before.

Note: the characteristic form of the osazones is a useful means of identifying sugars.

5.4 PREPARATION OF A 'SUPERSATURATED' SOLUTION

(crystallization (supersaturation))

Required

Potassium(I) sodium(I) 2,3-dihydroxybutanedioate (Rochelle salt); disodium(I) thiosulphate(VI); cotton wool

Procedure

1 Add small quantities of potassium(I) sodium(I) 2,3-dihydroxybutanedioate, with shaking, to a boiling tube about one-third filled with distilled water.

2 When no more solid will dissolve, add sufficient excess to form a 1 cm layer of the salt at the bottom of the tube.

3 Heat gently until the solid dissolves and then plug the tube with cotton wool and allow to cool in a rack. If any solid separates on cooling, reheat and cool once more.

4 The solution is now 'supersaturated' and crystallization can be initiated by scratching the inner surface of the tube with a glass rod or 'seeding' the solution with a crystal of potassium(I) sodium(I) 2,3-dihydroxybutanedioate.

5 One-quarter fill a boiling tube with disodium(I) thiosulphate(VI) crystals and heat gently.

6 Notice that the crystals dissolve in their own water of crystallization. Cool as before to form a 'supersaturated' solution. Seed with a single crystal of disodium(I) thiosulphate(VI) and note the effect.

6 Distillation

Distillation is commonly used in the purification of organic liquids. If only non-volatile impurities are present then the simple distillation process already described for boiling point determinations is effective (page 5). Should the impurities themselves be volatile liquids, a single distillation of this kind would do little to separate the components of the mixture, unless their boiling points were markedly different. In this case a tedious course of repeated distillations would be necessary to produce a complete separation.

Fractional distillation is in effect a technique for carrying out a whole series of such simple distillations in one continuous operation. The fractionating column provides a large heat exchanging surface with a gentle temperature gradient from top to bottom. At any point in the column there is equilibrium between the ascending vapour phase and descending liquid phase. The vapour reaching the top of the column is rich in the most volatile component, and the 'scrubbed' condensate dripping back into the flask consists mainly of the least volatile one.

Intimate contact must be maintained between vapour and condensate to produce a good degree of separation and this is obtained by increasing the surface area of the column by infolding or inserts. To be more effective, the column may be loosely packed with small pieces of glass, porcelain, stainless steel, or other inert material, allowing the condensate to trickle back through the crevices against the flow of vapour. The efficiency of fractionation rises with the reflux ratio, i.e. the ratio of condensate returned down through the column to that removed as distillate. Excessive or irregular cooling is prevented by surrounding the column with a heat insulator. An electrical heating tape may be necessary when distilling fractions with high boiling points. Complex 'molecular' stills with electrically heated jackets are used for the complete fractionation of high boiling point liquids.

A complication occurs with fractionating liquid systems which are not ideal, i.e. do not conform to Raoult's Law*. These form azeotropic (constant boiling) mixtures which behave as pure liquids and boil at a steady temperature, which is usually lower than that of either of the components (minimum boiling solution). With minimum boiling solutions the binary azeotrope distils off until one of the components has been removed completely. The temperature then rises and the component in excess boils off alone (fig 6.1).

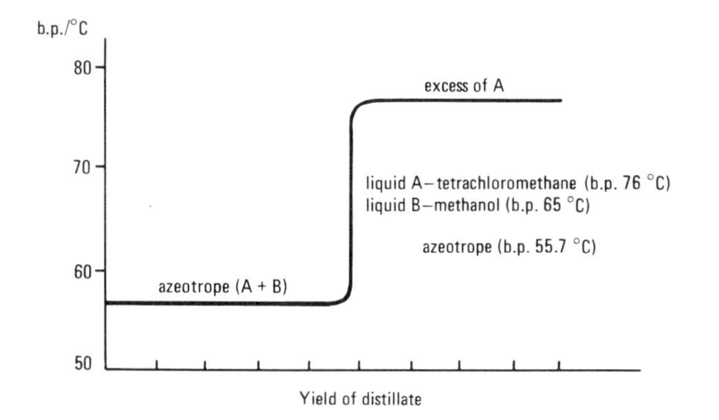

Fig 6.1 Distillation curve for minimum boiling mixture

* Raoult's Law states that for ideal dilute solutions at given temperature, the lowering of vapour pressure of the solvent is proportional to the mole fraction of the solute in the solution.

When the boiling point of an azeotrope is above that of either of the components (maximum boiling solution) the component in excess boils off before the azeotrope. Binary azeotropes can often be broken by the addition of a suitable third liquid. This technique is of great industrial value. Thus if benzene is added to an azeotropic mixture of ethanol and water (95% ethanol), fractionation gives a ternary distillate followed by an azeotropic mixture of benzene and ethanol and finally pure ethanol.

Experimental details

The simplest fractionating columns are of the type where the surface area is increased by folding or inclusions (pear or Vigreux columns). In other patterns the increase of surface is obtained by means of a central rod carrying circular discs, or in the Dufton column, by a flat glass spiral. These different types of column are shown in fig 6.2.

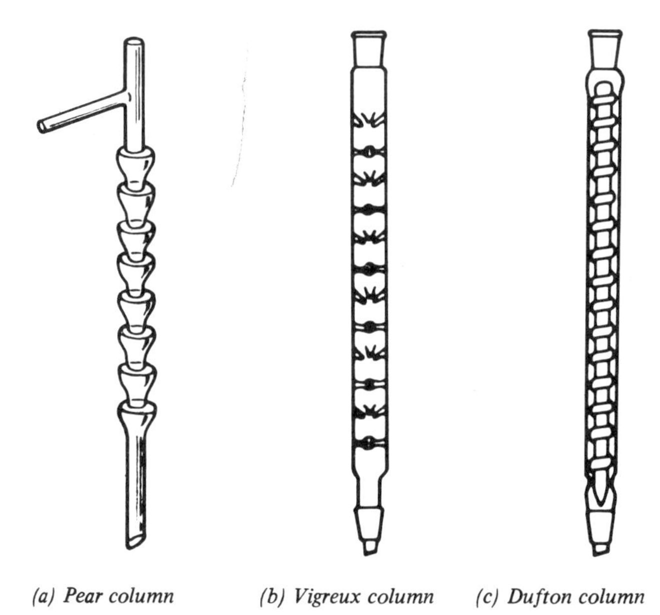

(a) Pear column *(b) Vigreux column* *(c) Dufton column*

Fig 6.2

For more efficient fractionation 'packed' columns are used in which an extremely large surface area is presented to the distillate vapour. A number of commercially produced column packings are available in the form of glass, stainless steel or porcelain rings, beads, and helices. To prevent these slipping through, the column usually has a fine glass bar welded across its lower end, or a slight constriction in the bore. Haphazard arrangement of the sections is necessary to prevent the condensate 'channelling'. This is best achieved by filling the column while in a horizontal position using a powder funnel, and then jerking it into an upright position. The more volatile the components to be separated the higher the packing should be extended up the column. For high boiling point mixtures the column can be insulated by wrapping with cotton or glass wool held in position with two or three pipe cleaners wound round at intervals.

A cheap and effective alternative packing is provided by a copper or stainless steel kitchen pot scourer, which can be bought at any hardware store. This can be pulled out into a long strand and carefully pushed into the column using a length of stout glass rod or wooden dowel. A small quantity can be tested for inertness by boiling

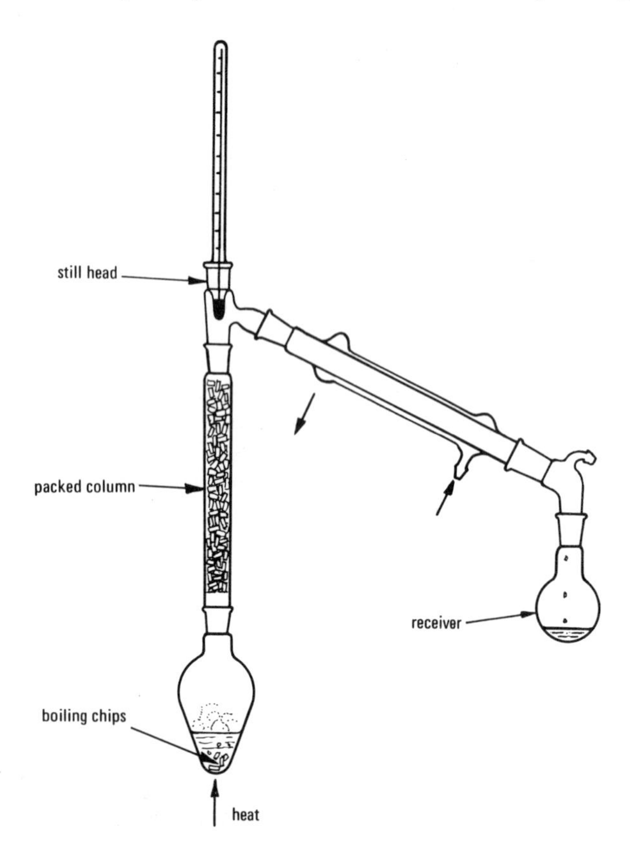

Fig 6.3 Semi-micro fractional distillation assembly

with a little of the mixture to be used before beginning distillation. A small scale fractionating column can be made by packing the neck of an ordinary distilling flask in this way.

The type of glass jointed assembly illustrated in fig 6.3 is most convenient for fractionating liquid mixtures. Rubber stoppers should never be used because they are affected by so many organic liquids. Care should be taken that the fractionating column is vertically above the distillation flask which should always contain a few boiling chips. Anti-bumping granules made of fused aluminium(III) oxide are very suitable for this purpose. It is helpful to collect the distillate in a small graduated flask if the volume of individual fractions is to be recorded.

The mixture is carefully warmed using a small bunsen flame or heating mantle set at low heat. The ring of condensate should slowly ascend the column, reaching the top after about ten minutes. During this period the heating should only be increased if the condensate level appears to be stationary. Once the column has warmed up the distillation should settle down to a rate of about a dozen drops of condensate a minute.

The distillation temperature should remain steady until the whole of the first fraction has been removed. At this point it will be necessary to raise the temperature of the mixture slowly until the appearance of the next fraction is indicated by a sharp rise in the thermometer reading. The secret of successful fractionation is patience, the whole process being carried out as slowly as possible.

The effectiveness of different types of fractionating columns in separating the components of a given liquid mixture can be seen by graphically recording the distillate yields over identical temperature ranges.

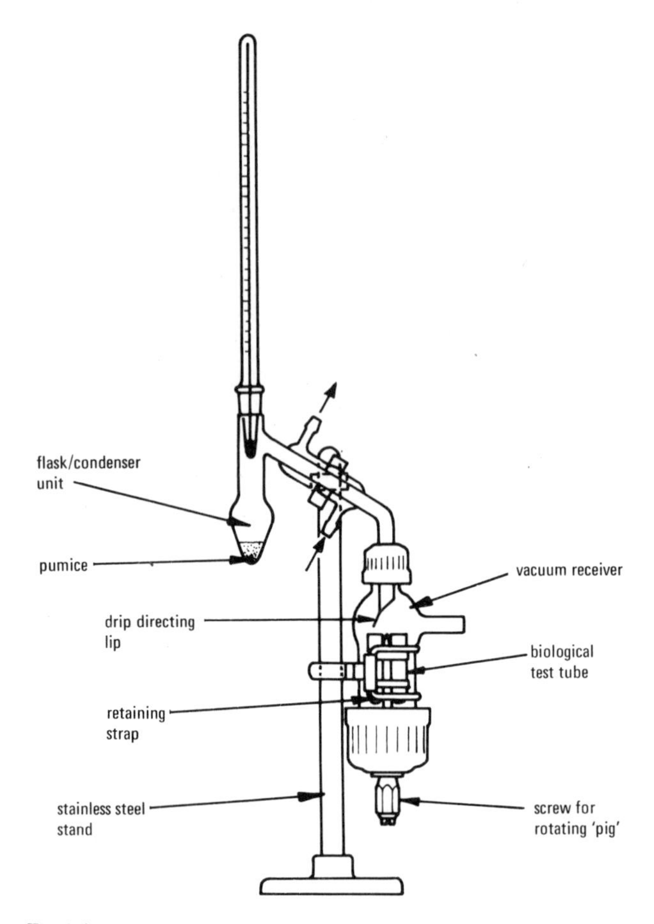

Fig 6.4 Micro distillation assembly

Micro-distillation

Occasionally synthesis involves the purification by distillation of very small quantities of volatile organic materials. For this purpose micro-distillation units (fig 6.4) are available with a capacity of 2–5 cm^3. A small pear-shaped distilling flask is used to minimize the decomposition of material by film formation on the glass surface. The water condenser is welded to the neck of the flask and passes into a vacuum receiver unit by means of a screw cap connector. Within the receiver unit are four standard biological test tubes which can be rotated on a central spindle to enable separate fractions to be collected without interrupting the distillation. This type of receiver is known as a 'pig'.

Heating is carried out by means of a micro-burner and a pinch of pumice powder or a short length of very fine capillary tubing is used to prevent bumping. A distillation rate of about 0.2–0.3 cm^3 a minute is desirable to minimize losses by thermal decomposition.

Distillation under reduced pressure

Many substances decompose below their boiling points at atmospheric pressure and cannot therefore be distilled in the normal way. Thus attempts to distil heavy oil fractions at ordinary pressure would inevitably result in

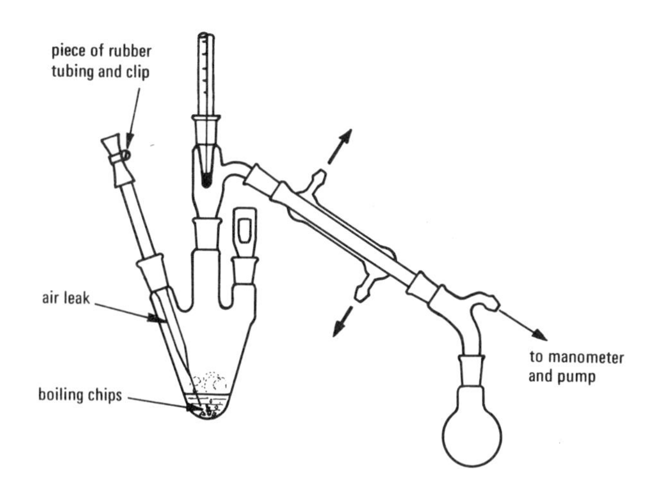

Fig 6.5 Low-pressure distillation assembly

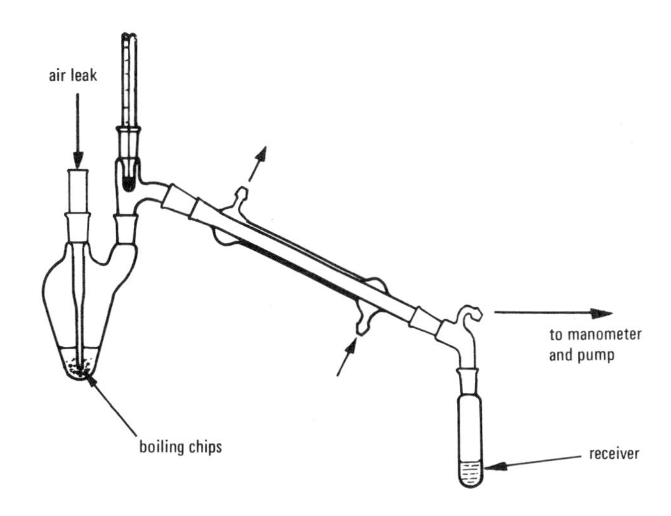

Fig 6.7 Distillation under reduced pressure using Claisen flask

thermal cracking of the product. In other cases the boiling point of a liquid is so high that distillation becomes difficult. By reducing the pressure during distillation, however, the boiling point can be considerably reduced and the operation carried out without fear of decomposition. A simple type of apparatus for carrying out distillation under reduced pressure in this way is illustrated in fig 6.5.

It will be noticed that a triple-necked flask is used to enable an 'air leak' to be fitted. When distilling at low pressure special precautions have to be taken to avoid bumping or frothing due to the liquid in the flask superheating. Here the use of boiling chips is not a sufficient safeguard and a short length of capillary tube drawn out at its lower end into a very fine jet is used. This allows a stream of very small bubbles to be drawn through the liquid during distillation. Superheating is thus minimized, the reduction of pressure due to the air admitted being negligible.

A simple manometer assembly for use when carrying out distillation at low pressure is illustrated in fig 6.6. The manometer, constructed from stout-walled glass tubing, is firmly fixed to a wooden support and dips into a small bottle of mercury. The end of a metre rule is sawn off at the 80 cm mark and a white line painted across at 76 cm acts as the zero. The rule is fixed alongside the manometer by means of two slots. A wing nut on the top of the slot enables the zero to be quickly adjusted to the level of mercury in the reservoir when taking a reading. By inverting the rule in this way the pressure in mm of mercury can be read off directly from the scale.

A specially designed twin-necked Claisen flask is also commonly used for distillation at reduced pressure (fig 6.7). This has a curved side arm which carries the thermometer and condenser unit and an upright neck into which is fitted the air leak.

A disadvantage of the assemblies described above for distillation under reduced pressure is that in order to remove each fraction the distillation has to be stopped and the vacuum broken. To avoid this either a revolving 'pig' can be used as in the micro-distillation assembly

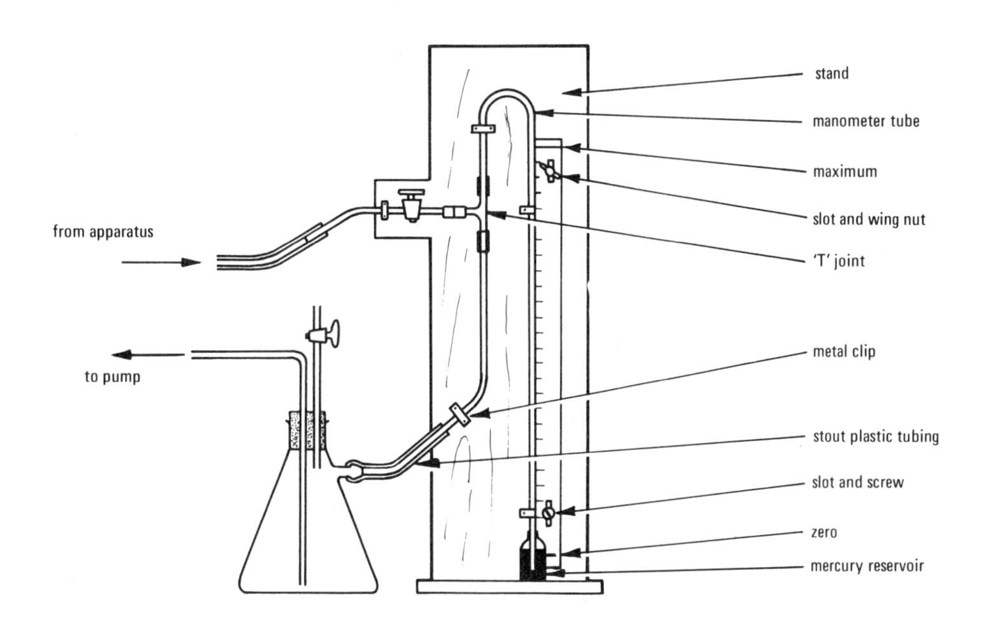

Fig 6.6 Manometer assembly for distillation at reduced pressure

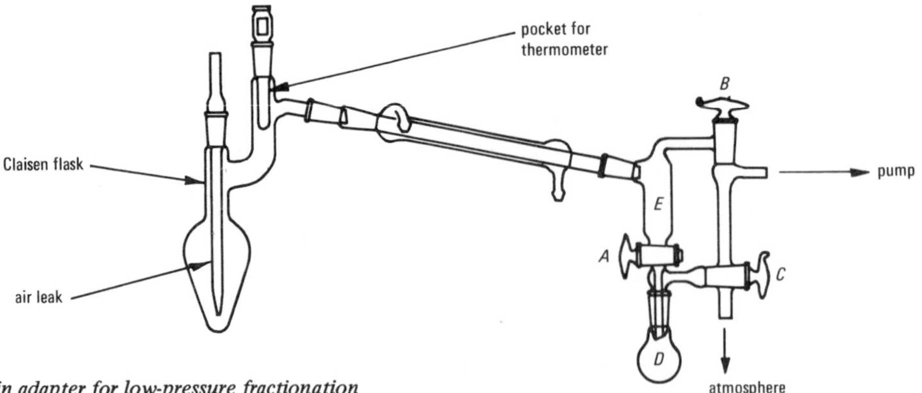

Fig 6.8 Use of Perkin adapter for low-pressure fractionation

fig 6.4) or a by-passing device known as a Perkin adapter (fig 6.8).

When distillation is started, A is opened and B and C turned to connect the two receivers D and E to the pump. As soon as the first fraction has passed into the receiver D, the tap A is closed and the second fraction collected in E. The first fraction can now be removed, the receiver being isolated from the pump by opening C to the atmosphere. On replacing D it can be reconnected to the pump and the second fraction run into it and removed as before. If the pump is of low power it is necessary to close B for a few seconds when replacing the receiver to allow the pressure to fall before opening A.

Distillation at reduced pressure can obviously only be carried out satisfactorily if the whole assembly is air-tight. It is an advantage, therefore, to use silicone-greased ground glass jointed apparatus if possible.

When distilling liquids with high boiling points it is often helpful to wind asbestos string round the neck of the flask up to the level of the condenser outlet. For insulating flasks on micro or semi-micro scale, lengths of burette cleaner (pipe-cleaners) can be used.

Steam distillation

Steam distillation is an ingenious alternative to distillation at reduced pressure. It may be used for the separation of liquids which are immiscible with water and of low volatility, at temperatures well below their boiling point. In the laboratory it is particularly useful for extracting a relatively small yield of product from a large bulk of tarry residue which would be difficult to separate by other means.

In a mixture of two immiscible liquid phases, each liquid vaporizes as if the other were not present. When the total combined vapour pressures are equal to the atmospheric pressure the mixture boils at a temperature below that of either of the pure components. In this way many organic liquids of high boiling point can be distilled at a temperature below the boiling point of water by injecting live steam into the crude product. For example octane, which boils at 125 °C, can be steam distilled at a temperature of 89.4 °C.

The amount of a particular component in the steam distillate mixture is proportional to its partial pressure at the boiling point. Thus the masses of the components present in the distillate can be obtained by multiplying their partial pressures by their molecular weights. This relation was used by F.E. Neumann in 1877 to determine the relative molecular masses of high boiling liquids such as methylbenzene.

Many organic compounds which are steam distilled are not completely insoluble in water at the temperature of distillation. This naturally reduces the efficiency of the separation, but can be partially offset by adding a little

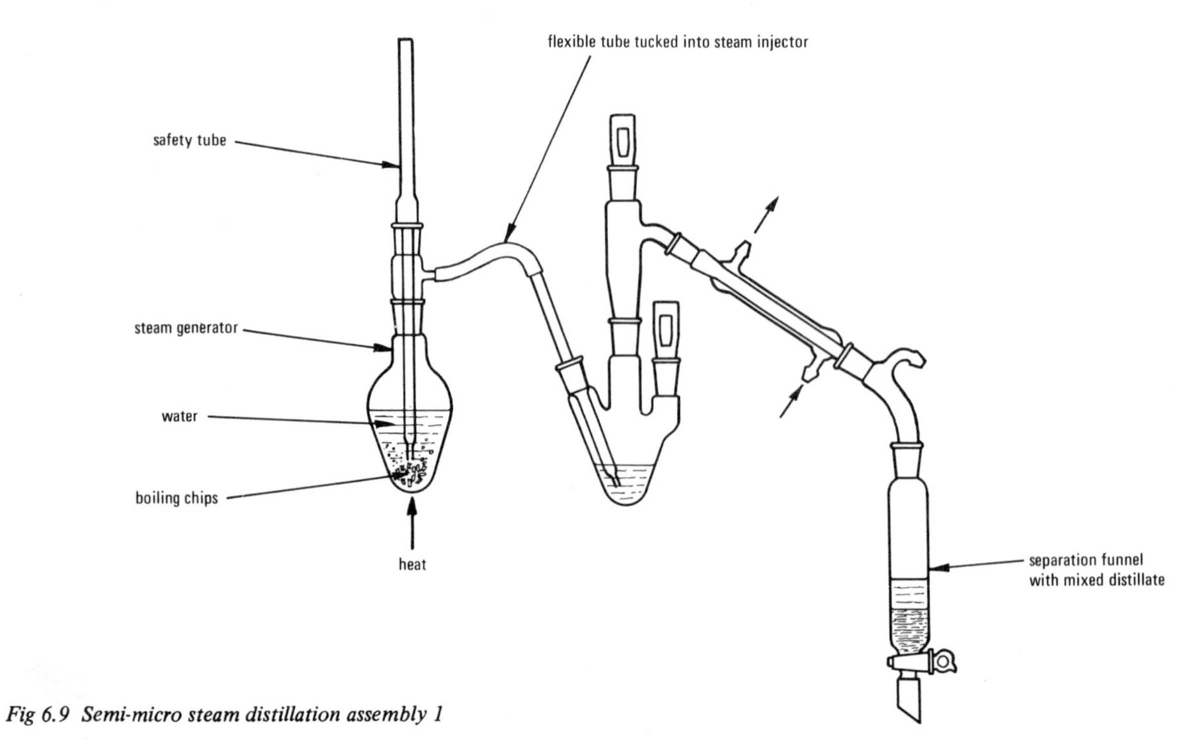

Fig 6.9 Semi-micro steam distillation assembly 1

sodium chloride to the mixture. The effect of this is to lower the vapour pressure of the water and decrease the solubility of the organic substance. A convenient semi-micro steam distillation assembly is illustrated in fig 6.9.

On this scale there is no need to heat the distillation flask, although it is necessary to do so when working on the macro scale to prevent undue condensation of steam. A pear-shaped flask acts as a steam generator as shown, the gas inlet tube being used as a safety valve. A short length of flexible plastics tubing is used to introduce the steam into the injector. The steam lead must be disconnected after the distillation has been completed to avoid sucking back. Where the quantity to be distilled is small, say less than 3 g, the steam generator can be dispensed with and distillation carried out by simply boiling the contents of the flask with a little water. The use of a separating funnel as a receiver facilitates the removal of the organic distillate from the aqueous layer.

A convenient 'all-in-one' assembly has been introduced for steam distillation on a semi-micro scale (fig 6.10). Here the distilling flask is inside the steam

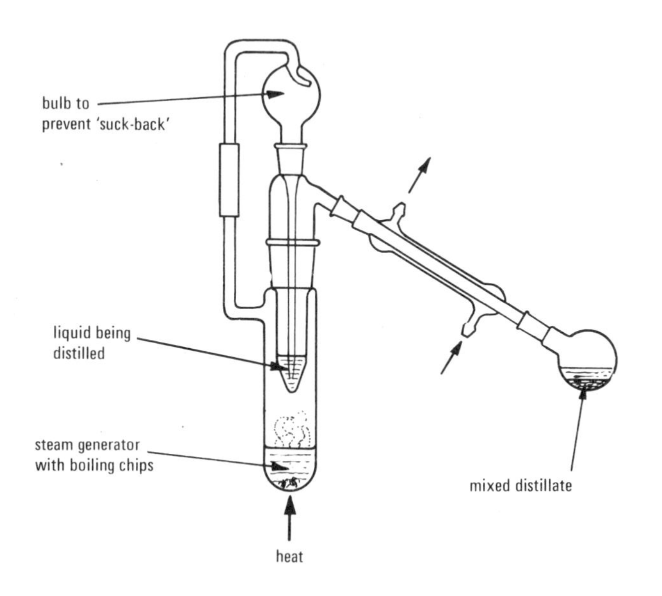

Fig 6.10 Semi-micro steam distillation assembly 2

generator thus keeping condensation to a minimum. Sucking back is prevented by means of a trap placed in the steam injector.

Occasionally during steam distillation the condenser becomes blocked with solid condensate. This can be removed by draining the water jacket of the condenser and allowing the steam to purge the tube.

Distillation on an industrial scale

Continuous distillation techniques are rarely required in the laboratory but are essential in industry. Extensive use of such processes is made in the fractionation of coal tar and crude petroleum, and in the rectification of a number of important organic reagents such as ethanol, phenylethene, and buta-1,3-diene.

The throughput of industrial fractionating columns is enormous and giant structures over 30 m high and 9 m in diameter are in common use. These are divided horizontally by a number of perforated metal plates called trays. Simple sieve plates are used occasionally when the feedstock contains solid matter — such as the alcoholic washes produced by mash fermentation. A more efficient

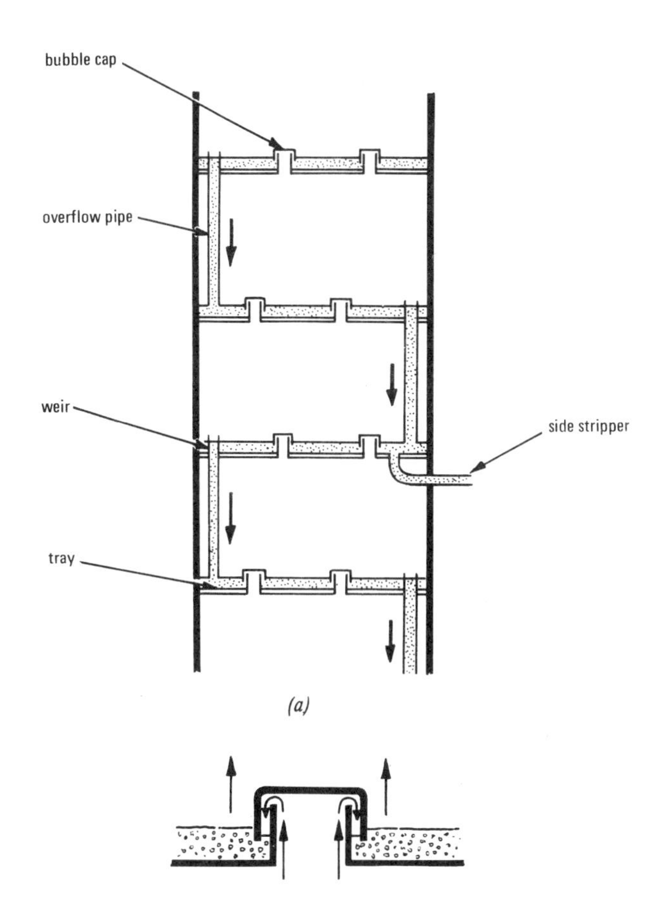

Fig 6.11 (a) Industrial fractionating column
(b) Detail of bubble cap

tray design incorporates bubble caps (fig 6.11). These are cylindrical metal caps with serrated lower edges which enclose the open tops of short vapour ports in the plates. The rising vapour initially condenses in each of the trays in turn until the liquid level just covers the base of the bubble caps. The vapour has now to bubble through the condensate in each tray as it rises, which ensures that it is thoroughly 'scrubbed'. In addition each plate is provided with a weir to enable the excess of condensate to overflow to the tray beneath which will be at a slightly higher temperature. The lighter fractions of this overflow will thus be re-evaporated. The height of the weir is adjusted to maintain a constant level of condensate in the trays. The bases of the bubble caps must not be too deeply submerged otherwise the ascending vapour would have to force its way through an appreciable head of liquid.

Feedstock from a simple pipe still is injected at a controlled temperature into the side of the column near to its base. The temperature gradient of the column is controlled by supplying additional heat to the liquid at the base of the column using a reboiler. The top of the column is kept at the required temperature by pumping back condensate at a controlled rate (refluxing). Each tray has a tendency to collect distillate of a narrow boiling point range. Inevitably a small percentage of liquid of lower boiling point than the bulk of the fractions present will be removed from the ascending vapour stream by 'entrainment'. This is removed by passing the cut from selected trays through a small subsidiary column called a stripper. The liquid is allowed to pass down through bubble trays against an upward current of steam. This strips off the more volatile fraction which is returned to the column with the steam.

Unit for the fractional distillation of chlorosilane feedstock during the manufacture of silicones

The cost of fractionation lies mostly in the heating of the still and the subsequent cooling of the distillate. A saving in fuel costs has been made by the installation of heat exchangers in which the crude feedstock can be preheated by both the outgoing residue and the hot condensing vapours. The heat exchanger is essentially a system of tubes which carry the cold feed, these being immersed in a current of hot liquid or vapour from the column. The preheated feed passes into the furnace of a pipe still where further heating takes place before it is injected into the column.

The efficiency of a fractionating column is expressed in terms of the number of 'theoretical plates' it contains. This is a purely hypothetical figure which is obtained by comparison of the composition of the liquid at the top and bottom of the column (or section). Thus a theoretical plate in a distillation column may be defined as a plate holding a pool of liquid, the vapour rising from which is in equilibrium with the liquid leaving the plate. The ratio of the number of theoretical plates to the number of actual plates present represents the efficiency of the column which varies from about 10% to over 100%. The height of the column divided by the number of theoretical plates is termed the 'HTU' or 'height of theoretical unit'.

Extractive distillation is occasionally used to separate a component from a mixture being distilled. As in azeotropic distillation this is done by adding another liquid component. Here, the added liquid, which is of high boiling point, dissolves the required extractive and reduces its vapour pressure. The solution is carried down the column and finally becomes a bottom product. The required material is recovered from this by stripping in a secondary column. A typical example of extractive distillation is the recovery of methylbenzene from a mixture of naphthenates, metal salts of carboxylic acids derived from substituted cyclohexanes, and alkanes, by pumping phenol into the top of the column.

Distillation at reduced pressure is often used industrially for the removal of water from foodstuffs—as in the preparation of condensed and evaporated milk. This enables concentration to be effected at low temperatures, minimizing protein and vitamin damage. The saving in fuel, however, if offset by the expensive vacuum equipment required. In addition, the increase in the volume of vapour produced at low pressure (Boyle's Law) requires a corresponding increase in distillation capacity. Increasing use is being made of freeze-drying in this field (see page 3).

Steam distillation is also used on an industrial scale. Since water has a low molecular weight economic separations can be carried out with liquids of low volatility as long as they have a compensating high molecular weight.

For example, benzenecarbaldehyde prepared industrially by liquid phase oxidation of methylbenzene is extracted by steam distillation. The mixture boils at 98 °C (b.p. benzenecarbaldehyde, 180 °C) and the mixed distillate contains benzenecarbaldehyde and water in a mass ratio of roughly 1:2.

Distillation experiments

6.1 RECOVERY OF DRY CLEANING FLUID

(simple distillation)

Required

Thawpit (tetrachloromethane); domestic paraffin; ethanol (industrial methylated spirit)

Piece of dirty woollen cloth*; broken porcelain or boiling chips

Procedure

1 Shake the piece of dirty wool with about 10 cm^3 of tetrachloromethane in a stoppered separating funnel.

2 Decant off the impure solvent and distil using the apparatus shown in fig 2.1.

3 Record the boiling range of the distillate.

Sequel

Repeat the experiment using samples of coloured domestic paraffin and ethanol coloured with ballpoint ink.

6.2 COMPARISON OF DISTILLATION ASSEMBLIES

(fractional distillation: Dufton column; packed column; pear column)

This is a convenient experiment for a group of students, each carrying out the separation with a different assembly.

Required

Propanone; broken porcelain or boiling chips

Procedure

1 Fractionate a 1:1 mixture of propanone and water using a simple distillation unit (fig 6.3).

2 Half fill the pear flask with the mixture and add three or four pieces of porcelain.

3 Label and number five small snap-top glass containers (or small stoppered flasks).

4 Read the hints on distillation on page 21-2 and collect fractions boiling in the following temperature ranges: (a) 56-62 °C, (b) 63-72 °C, (c) 73-82 °C, (d) 83-95 °C, (e) residue.

5 Change the receivers quickly at the specified temperature and after collecting (d) allow the apparatus to cool and pour the contents of the distilling flask into (e).

6 Measure and record the volume of each fraction [(a) and (e) should be almost pure propanone and water respectively)].

7 Draw a distillation curve plotting volume of distillate against temperature.

*Wet the cloth with a drop of bicycle oil and wipe it on the floor.

Sequel

1 Repeat the experiment using any other type of fractionating column available (page 21).

2 Superimpose the distillation curves obtained and comment on the efficiencies of the columns concerned.

6.3 FRACTIONAL DISTILLATION (MICRO SCALE)

(micro-distillation unit; 'pot scourer' column)

Required

Methylbenzene; tetrachloromethane; pumice dust

Procedure

1 Remove the thermometer of the micro-assembly (fig 6.4) and half fill the flask with a 1:1 mixture of methylbenzene and tetrachloromethane. Add a pinch of pumice dust to prevent 'bumping'.

2 Pack the neck of the flask with a short length of unravelled copper pot scourer.

3 Replace the thermometer and turn the screw so that one of the receiver tubes is below the condenser outlet. Make sure that water is flowing in the condenser.

4 Gently heat the flask and collect the fractions boiling over the following temperature ranges: (a) 76–82 °C, (b) 83–92 °C, (c) 93–102 °C, (d) 103–112 °C, (e) residue.

5 Turn the receiver to collect each of the first four fractions in turn, the residue being the liquid remaining in the flask.

6 Measure and record the volume of each fraction.

7 Draw a distillation curve plotting fraction volume against temperature.

6.4 FRACTIONATION OF CRUDE OIL MIXTURE

(fractional distillation at reduced pressure)

Required

Petroleum ether (b.p. 40–60 °C);
petroleum ether (b.p. 60–80 °C); synthetic turpentine; kerosine; bicycle oil; bitumastic paint; paraffin wax

Special equipment: assembly for fractional distillation at reduced pressure; heating mantle.

Procedure

1 Make up a synthetic 'crude' by shaking together the following:

petroleum ether (b.p. 40–60 °C)	10%
petroleum ether (b.p. 60–80 °C)	10%
synthetic turpentine	20%
kerosine (paraffin)	20%
bicycle oil	20%
bitumastic paint	15%
paraffin wax	5%

2 One-third fill a twin-necked flask with the 'crude' and attach a 'pot scourer' fractionating column insulated with asbestos tape and fitted with a water condenser.

3 Add a few anti-bumping granules to the flask and heat with an electric mantle.

4 Collect fractions at 20 °C intervals and keep in labelled stoppered tubes. (Empty the water condenser when the distillate temperature reaches about 130 °C).

5 When the temperature of the distillate approaches 200 °C stop heating and arrange the assembly for vacuum distillation, replacing the stopper in the distillation flask by an air leak.

6 Turn the filter pump on full and continue heating, collecting labelled fractions as before until the contents of the flask show signs of decomposing.

7 Disconnect the pump, allow the flask to cool somewhat, and then pour the contents into a container labelled 'residue'.

8 Examine the fractions and record their colour, smell, and volume.

Note: samples of crude petroleum which have been 'topped' to remove the more volatile fractions can be obtained by schools from the Shell International Petroleum Co. Ltd, Shell Centre, London SE1 7NA.

Sequel

Suggest a laboratory scale process for reducing the bulk of milk by evaporation. What special problems would have to be overcome?

6.6 DEHYDRATION OF ETHANOL USING AN AZEOTROPIC TECHNIQUE

(fractional distillation)

Required

Ethanol (industrial methylated spirit); water, benzene*

Broken porcelain; labelled snap-top bottles

Procedure

1 Use the packed column distillation assembly in fig 6.3.

2 Prepare an ethanol/water/benzene mixture in a ratio of 9:1:2. Half fill the distilling flask with the mixture and add a few pieces of broken porcelain.

3 Distil the ternary mixture, collecting fractions in labelled bottles.

4 Record the yield and boiling point range of each fraction.

***CARE: danger from skin absorption and inhalation.**

6.5 EXTRACTION OF SUGAR FROM BEET

(evaporation at reduced pressure)

Required

Sugar beet; calcium(II) hydroxide; decolorizing charcoal

Filter pump; broken porcelain or boiling chips

Procedure

1 Grate some scrubbed beet into a 250 cm^3 beaker (use a kitchen grater) and cover well with water.

2 Heat to 80 °C, keep at this temperature for about 10 minutes, then cool.

3 Filter off the aqueous extract using a large funnel containing a plug of glass wool and pressing down the pulp with an inverted glass stopper.

4 Add a little calcium(II) hydroxide and decolorizing charcoal and boil for a few moments. Then add some filter pulp and filter through a Buchner funnel.

5 Cool and then bubble carbon dioxide gas through the purified extract (gas generator or cylinder) for five minutes to decompose the calcium(II) sucrosate.

6 Boil the sugar solution in a beaker until the volume has been reduced by half.

7 Using an assembly for distillation at reduced pressure (fig 6.5) remove the bulk of the remaining water by distillation with the pump at full suction.

8 Pour the hot concentrate into a dish, cover with a watch glass, and leave to crystallize.

9 Filter off the sugar crystals using a Hirsch funnel and examine their form under low power magnification.

6.7 ISOLATION OF NAPHTHALENE FROM ADULTERATED SAMPLE (MOTHBALL)

(steam distillation, small scale)

Required

One mothball; soot; sugar

Procedure

1 Crush the mothball in a mortar and mix with a little soot and sugar.

2 One-quarter fill the inner distillation tube of the apparatus shown in fig 6.10 with the impure powdered naphthalene and cover with a little water.

3 Heat the boiling tube so that a steady stream of steam passes into the mixture. (Note: if during the distillation the condenser becomes clogged with solid, turn off the water for a few moments to enable it to melt away.)

4 When a reasonable sample of naphthalene has collected, disconnect the receiver, and scrape out the contents onto a wad of three or four filter papers.

5 After drying the naphthalene by pressing on the filter papers check its melting point (naphthalene m.p. 80 °C).

Sequel

Assume that the boiling point of a mixture of naphthalene and water is 99 °C at 760 mmHg pressure and that at this temperature the respective vapour pressures of naphthalene and water are 18 mmHg and 742 mmHg. Calculate the proportions of naphthalene and water which the mixed distillate should contain (relative molecular mass naphthalene 128).

6.8 STEAM DISTILLATION OF A MIXTURE OF 2-HYDROXYBENZOIC ACID AND 1,4-DICHLOROBENZENE

(steam distillation, semi-micro scale)

Required

2-hydroxybenzoic acid; 1,4-dichlorobenzene; hydrochloric acid

Procedure

1 Grind in a mortar a mixture of two parts of 2-hydroxybenzoic acid to one part of 1,4-dichlorobenzene.

2 One-quarter fill the three-necked flask of the steam distillation unit (fig 6.9) and add an equal amount of water. Replace the tap funnel receiver shown in the diagram with a small conical flask.

3 Steam distil the mixture, occasionally running water through the condenser, if necessary, to condense the steam. If the condenser becomes choked, drain off the water altogether until the blockage clears.

4 Continue distillation until the distillate is clear. Filter off the solid distillate using a Hirsch funnel (fig 3.2a) and dry between filter papers.

5 Cool the residue in the distillation vessel, place in a small conical flask and add 2 cm^3 of concentrated hydrochloric acid. Filter off the solid through the Hirsch funnel and dry as before.

Sequel

1 Determine the melting points of the original mixture, the distillate and residue (2-hydroxybenzoic acid m.p. 156 °C; 1,4-dichlorobenzene m.p. 53 °C).

2 How effective is steam distillation in separating the two components of the mixture?

6.9 STEAM DISTILLATION OF A MIXTURE OF BENZENE AND TECHNICAL XYLENE

(steam distillation, semi-micro scale)

Required

Benzene* (b.p. 80 °C); technical xylene (mixture of dimethylbenzenes b.p. 135 °C)

Two labelled bottles

Procedure

1 Make a mixture of 4 cm^3 of water with 3 cm^3 each of benzene and xylene.

2 Steam distil the mixture using the apparatus shown in fig 6.9. Mark the receiver with a wax pencil at the 5 cm^3 level.

3 Collect two separate 5 cm^3 fractions of distillate in the receiver and pour into labelled bottles.

4 Measure and record the upper hydrocarbon layer in each 5 cm^3 sample.

Sequel

Predict the possible composition of the two distillate fractions, giving reasons.

***CARE: danger from skin absorption and inhalation.**

7 Solvent extraction

Solvent extraction is commonly used in organic chemistry for the separation of a liquid or solid from an aqueous suspension or solution. It is usually carried out in the laboratory by shaking the aqueous mixture in a separating funnel with an organic solvent which is immiscible with water, and then allowing the liquid layers to separate out. The components of the mixture distribute themselves between the two solvents in concentrations which are proportional to their relative solubilities. At any given temperature the ratio of these concentrations for a given solute is known as the distribution coefficient.

Inorganic salts which may be present as impurities will appear almost exclusively in the aqueous layer. On the other hand, hydrocarbons and their halides, which do not form hydrogen bonds, are virtually insoluble in water but readily soluble in most organic extraction solvents. Compounds of this type can often be extracted effectively in a single operation.

Hydrogen bonded substances such as ketones, aldehydes, esters, alcohols, acids, and amines are less easy to extract. The mixture must be repeatedly treated with a solvent in which the organic solute is considerably more soluble than in water.

With a given volume of solvent, the effectiveness of the separation achieved increases with the number of extractions performed. It is usual therefore to carry out three or four extractions with small volumes of solvent rather than a single extraction with a larger amount.

After separation from the aqueous layer the extracts containing the organic solute are dried by shaking with a suitable drying agent. The solvent is then distilled off, and the residue purified by crystallization, or if the extract is a liquid, by distillation.

The liquid most commonly used for extraction purposes is probably ethoxyethane. Apart from its powerful solvent properties, it has a low boiling point (35 °C) which simplifies its removal after the extraction has been carried out. The fire hazard involved in the use of ethoxyethane can be considerably reduced by careful handling and sensible precautions such as extinguishing flames in the vicinity of the separation. Other extraction solvents of importance are petroleum ether, benzene, trichloromethane, tetrachloromethane and chloromethane.

Experimental details

The separating funnel used should have a well fitting ground-glass stopper and stopcock. Before use these should both be wiped with a soft cloth and thinly smeared with petroleum jelly or silicone grease. The funnel should not be more than half-filled with the aqueous mixture and about one-third this quantity of solvent added. The funnel is then stoppered and inverted so that the tap section is uppermost. After shaking gently for a few seconds, taking care to hold the stopper in place with the thumb, the tap is opened cautiously to release any excess gas pressure which may have built up (venting) and then reclosed. After repeating this process the funnel is shaken vigorously for about two minutes and after venting once more it is placed stem downwards in a retort ring padded with short lengths of slit bunsen tubing (fig 7.1).

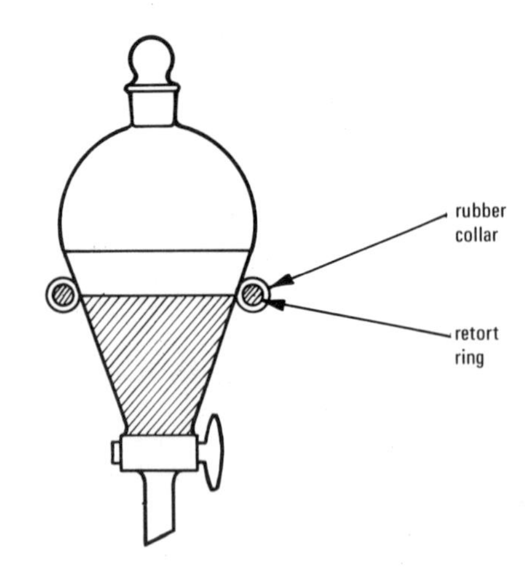

(a) Pear-shaped separating funnel

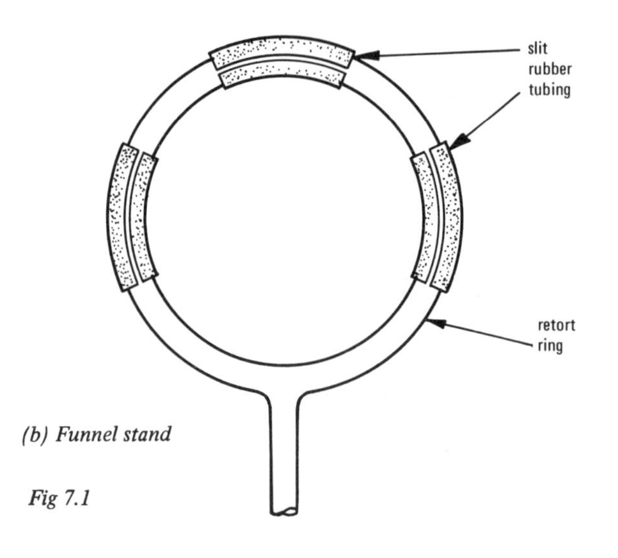

(b) Funnel stand

Fig 7.1

When the liquid has separated out into two distinct layers, the lower layer is run off after removing the stopper. The small amount of liquid at the solvent interface is discarded, and it is best to pour the upper layer from the top of the funnel to avoid contamination via the stem and tap section. The organic extract is retained in a stoppered flask while the aqueous layer is returned to the funnel and treated with a fresh quantity of solvent as before. Which of the two layers in the funnel is the aqueous one naturally depends upon the density of the other solvent used.

After three or four extractions the aqueous layer is discarded. (It is wise to retain the discard liquor if there is any doubt as to which layer is which!) The extracted fractions are now combined and shaken with a drying agent in a stoppered conical flask to remove traces of water. Anhydrous calcium(II) chloride is commonly used except with alcohols, phenols, or amines. After filtering off the drying agent the solvent is removed by distillation. If ethoxyethane has been used this operation should be carried out in a well ventilated fume cupboard using an electric heating mantle, or with small quantities, by heating the extract in a small beaker held in a bath of hot water.

When working with alkaline aqueous solutions an emulsion is sometimes formed which prevents separation of the two layers. The emulsion can occasionally be broken up by merely swirling or stirring with a glass rod, but centrifugation is often necessary. Another effective method of breaking the emulsion is to saturate the aqueous layer with salt — a process known as 'salting out'. This has the double advantage of reducing the solubility in the aqueous layer of both the organic solute and the solvent.

Extraction by salt formation

Basic or acidic organic compounds are often best separated from mixtures by conversion into water soluble salts which are insoluble in ethoxyethane. For instance sodium hydroxide solution can be used to convert carboxylic acids or phenols to their sodium salts which can then be removed in aqueous solution. The salts can be readily reconverted into the parent compound by acidifying with dilute hydrochloric acid. An aqueous solution of sodium hydrogencarbonate can be used in place of sodium hydroxide solution, except that it is not sufficiently alkaline to react with phenols. This provides a useful method of separating mixtures of phenols and organic acids.

Basic organic compounds, such as amines, can similarly be converted into water soluble salts by the use of dilute hydrochloric acid. In this case the aqueous solution is treated with sodium hydroxide solution to recover the parent base.

Continuous extraction

When an organic solute is more soluble in water than in the extracting solvent, or when emulsions are formed during extraction which are difficult to break, it is convenient to use a continuous extraction process. For this purpose a specially designed piece of apparatus is used in which a stream of small droplets of solvent is passed through the aqueous layer containing the solute. With a solvent such as ethoxyethane which is less dense than water, the solvent droplets collect as an upper layer above the aqueous phase (fig 7.2a). This upper layer is siphoned off continuously and heated, the refluxing droplets of pure solvent being recycled. In this way the solute gradually accumulates in the 'boil-up' vessel. For solvents denser than water the apparatus is modified to allow the solvent to fall through the aqueous phase (fig 7.2b). For small extractors where the circulation is fairly rapid a simple funnel distributor is adequate to distribute the solvent. In larger models, distribution is carried out either by baffles or by spreading the solvent into a spray of droplets through a sintered glass plate.

It is often necessary to carry out exhaustive and continuous extraction of solid samples with a hot solvent. Here it is usual to employ a Soxhlet extractor. The gently boiling solvent refluxes as a stream of drops on the solid which is contained in a small thimble of moulded filter paper. When the section of the Soxhlet which contains the thimble becomes filled with liquid it automatically siphons back into the boiling solvent. This cycle is repeated until extraction is complete. The solute may then

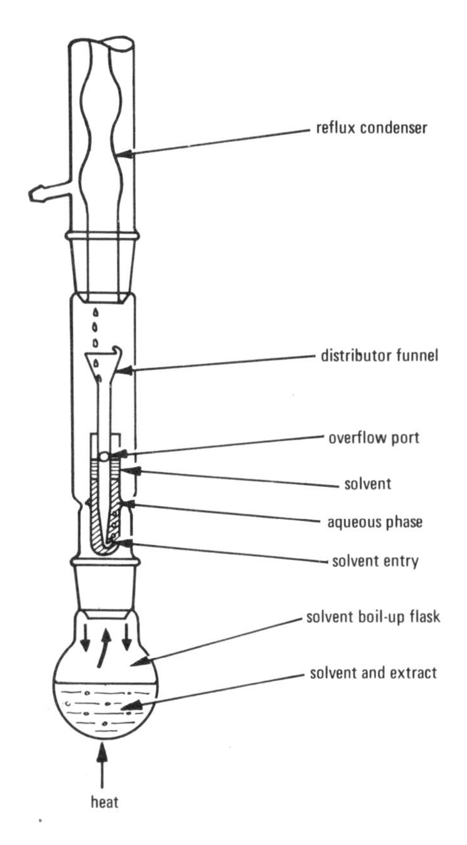

(a) upward displacement type

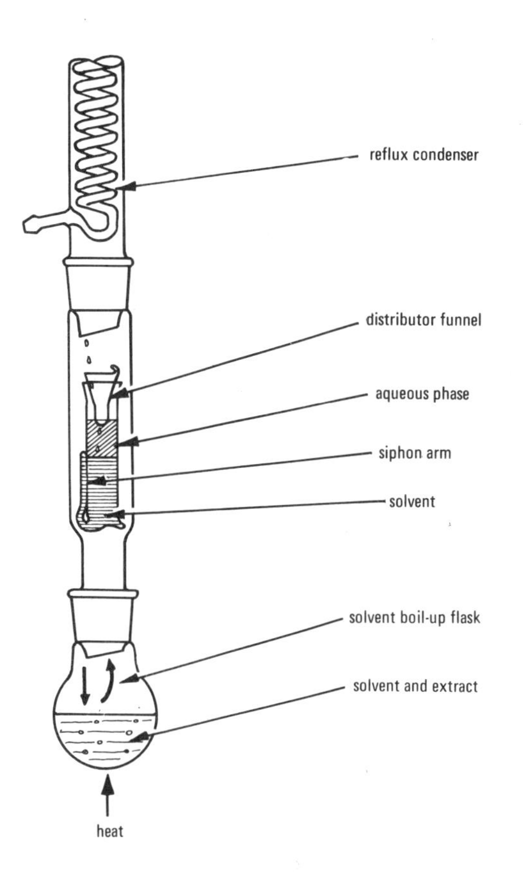

(b) downward displacement type

Fig 7.2 Semi-micro scale continuous extraction apparatus

be isolated by distilling off the solvent.

An improved version of the Soxhlet has the thimble tube and siphon surrounded by a jacket of hot solvent vapour (fig 7.3). This enables the extraction to be carried out at a higher temperature and lessens the likelihood of damage to the delicate siphon tube. The isolation of organic substances from plant and animal tissues having a high water content is conveniently carried out in this type of apparatus.

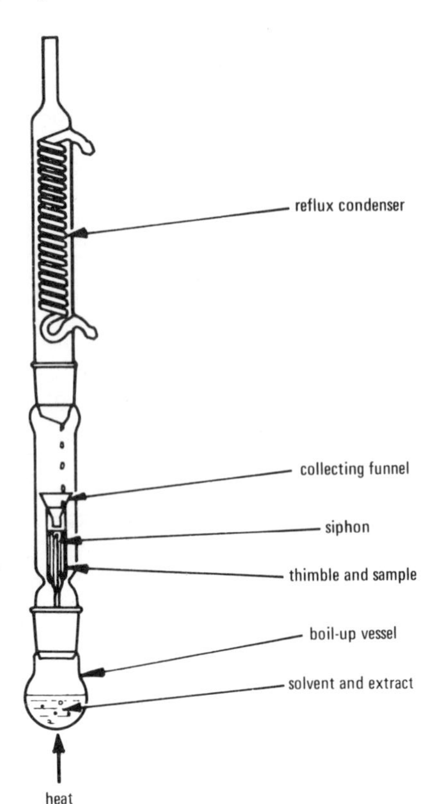

Fig 7.3 Enclosed-type Soxhlet extractor

Drying organic liquids

Organic liquids can conveniently be dried by treatment with small quantities of inorganic drying agents (table C). Care must be taken to choose a substance which, while being rapid and effective in its drying action, does not dissolve in the liquid or react with it chemically. The drying agent can be readily removed after use by filtration. Most of the common drying agents are the anhydrous forms of metal salts such as calcium(II) chloride, potassium carbonate and the sulphates of magnesium, calcium, and sodium. Quicklime (calcium(II) oxide) and alumina (aluminium(III) oxide) are also used, and for a high degree of desiccation, phosphorus(V) oxide. Metallic sodium is employed to remove traces of water from ethers and saturated hydrocarbons, but extreme caution is required in its use because of its great reactivity.

The liquid to be dried is shaken with small quantities of the drying agent in a stoppered conical flask. The used desiccant is removed by filtering through a fluted paper. When using calcium(II) chloride sufficient water may be present to form an aqueous layer which can be removed in a separating funnel prior to filtration. Ideally the liquid should be allowed to stand overnight in contact with the drying agent, and then distilled.

Liquids of high boiling point which are immiscible with

Table C – Common inorganic drying agents

Agent	Notes
anhydrous calcium(II) chloride	**not** to be used for alcohols, phenols, amines or acidic liquids
anhydrous sodium sulphate(VI) anhydrous magnesium(II) sulphate(VI) anhydrous calcium(II) sulphate(VI)	can be used for all liquids but are relatively slow
sodium or potassium hydroxide	not to be used for acids, phenols or esters; very suitable for amines
calcium(II) oxide	may be used for bases and alcohols
phosphorus(V) oxide	useful for ethers, saturated and aromatic hydrocarbons, and alkyl and aryl halides
anhydrous potassium carbonate	useful for ketones and amines

Columns used in the purification of lubricating oil by solvent extraction using furfural

water can be dried effectively by the addition of benzene and subsequent distillation of the resulting ternary mixture (page 21). Organic liquids are also occasionally freeze-dried, the water being frozen and the resulting ice sublimed off at low pressure.

Industrial applications

Solvent extraction is used industrially to remove specific components from mixtures when this would be difficult or impossible by other means. As early as 1843, attempts were made to extract vegetable oils from plant material using solvents. Currently oil cake is stripped of the 5% or so of oil remaining after expression using solvents such as hexane, petroleum ether, or trichloroethane. Normally the cake is carried on a moving belt countercurrent to a succession of solvent sprays, the fresh solvent being used to wash the most highly extracted material.

In the perfumery industry, solvent extraction is used to extract floral oils which would be decomposed by distillation. Petroleum ether is commonly used to produce an initial 'concrete' and this is further refined by extraction with ethanol to produce an 'absolute'. Liquefied butane under pressure and at a temperature of −15 °C is used in the Butaflor process as a solvent for extracting the most delicate oils such as lilac and gardenia.

Solvent extraction is widely used in the refining of petroleum products. In 1907, Edeleanu developed a process for removing aromatic hydrocarbons from kerosine using liquid sulphur dioxide as a solvent. This process is still in use, the lower density kerosine passing up a packed column countercurrent to a descending stream of denser liquid suphur dioxide. Furfural extraction is used in the petroleum industry to refine lubricating oils, liquid propane being employed to remove the bitumen from heavier oils.

A combination of distillation and solvent extraction is used to recover aromatics such as benzene and methylbenzenes from gasoline fractions. Phenols can similarly be recovered from coal tars produced during gasification processes.

Solvent extraction experiments

7.1 COMPARISON OF SINGLE AND MULTIPLE EXTRACTIONS

(solvent extraction)

Required

Crystal violet (or any water soluble dye); trichloromethane

Procedure

1 Dissolve a small fragment of crystal violet in 50 cm^3 of water by shaking in a stoppered 100 cm^3 conical flask.

2 Grease the taps and stoppers of two pear-shaped separating funnels and pour half the coloured solution into each.

3 To one funnel add 30 cm^3 of trichloromethane and shake vigorously, occasionally opening the tap to reduce any build-up of pressure (page 30).

4 After two minutes shaking, place the funnel in a retort ring padded by lengths of slit bunsen tubing and after removing the stopper allow the contents to separate out. Run the lower trichloromethane layer into a boiling tube.

5 Extract the second 25 cm^3 of the coloured solution with three successive 10 cm^3 portions of trichloromethane, taking the same precautions as before. Run off the three trichloromethane fractions into a second boiling tube.

6 Pour the extracted aqueous solutions from the tops of the separating funnels into another pair of boiling tubes.

7 Compare the intensity of colour in the two aqueous solutions and the two trichloromethane extracts and hence assess the efficiency of the two types of extractions used.

7.2 ESTIMATION OF DISTRIBUTION COEFFICIENT BY A COLORIMETRIC METHOD

(solvent extraction)

Required

Standard aqueous solutions of 2-hydroxybenzoic acid as follows:

S — 0.2 g in 100 cm^3 of distilled water

S1, S2, S3 — 1 cm^3 of S diluted respectively to 100, 200 and 300 cm^3 with distilled water

S4 — 2 cm^3 of S diluted to 100 cm^3 with distilled water

S5 — 3 cm^3 of S diluted to 100 cm^3 with distilled water

Butyl ethanoate;
freshly prepared iron(III) chloride solution

Procedure

1 Shake 5 cm^3 of S with an equal volume of butyl ethanoate in a pear-shaped separating funnel.

2 Allow the layers to separate, run off the lower aqueous layer into a labelled test tube, and mark it E1.

3 Repeat the operation using 5 cm^3 of S with 10 cm^3 of butyl ethanoate and label the aqueous extract E2. (Do not discard the butyl ethanoate fractions but pour them into a 'residues' bottle so that the ester can be recovered by distillation.)

4 Shake 5 cm^3 of S with an equal volume of butyl ethanoate, run off the two layers as before and return the aqueous extract to the funnel and shake with a further 5 cm^3 of solvent. Label the doubly extracted aqueous portion E3.

5 Into six test tubes labelled S–S5 pipette 1 cm^3 of each of the standard 2-hydroxybenzoic acid solutions followed by one drop of iron(III) chloride solution.

6 Add one drop of iron(III) chloride solution to E1, E2 and E3 and compare the colour with the standard tubes. Record the number of the tube most closely matched.

7 Comment on your results.

7.3 SEPARATION OF BENZENECARBOXYLIC ACID FROM 1,4-DICHLOROBENZENE

(extraction by salt formation)

Required

Benzenecarboxylic acid; 1,4-dichlorobenzene; ethoxyethane; sodium hydroxide solution (10%); calcium(II) chloride; dilute hydrochloric acid; naphthalen-2-ol

Procedure

1 Dissolve about 0.5 g each of 1,4-dichlorobenzene and benzenecarboxylic acid in 30 cm^3 of ethoxyethane (extinguish all flames) by shaking gently in a small conical flask.

2 Pour the solution into a pear-shaped separating funnel, add half this volume of 10% sodium hydroxide solution and shake, taking the usual precautions to release any pressure build-up (page 30).

3 Allow the two layers to separate and run off the lower aqueous layer into a small conical flask.

4 Pour the ether layer from the top of the funnel into a second flask and shake with a little calcium(II) chloride until it is no longer turbid. Decant off the ethoxyethane extract into a large watch glass and allow the solvent to evaporate (use a fume cupboard).

5 Add an equal volume of dilute hydrochloric acid to the aqueous solution and shake vigorously. Filter off the precipitated benzenecarboxylic acid using a Buchner filter and dry between filter papers.

6 Check the melting points of the two separated solids (1,4-dichlorobenzene m.p. 53 °C; benzenecarboxylic acid m.p. 123 °C).

Sequel

1 Naphthalen-2-ol and benzenecarboxylic acid both have a melting point of 123 °C. How would you separate a mixture of these two substances and 1,4-dichlorobenzene?

2 When you have devised a scheme try it out and check the melting points of the separated naphthalen-2-ol and benzenecarboxylic acid (page 2). Note: for this experiment equal masses of naphthalen-2-ol, 1,4-dichlorobenzene, and benzenecarboxylic acid can be melted together on a steam bath (how is this possible when two of the components melt at 123 °C?). The fused mixture is then ground up on cooling.

3 Supposing phenylamine was substituted for benzenecarboxylic acid in the foregoing experiment, how could the components of this mixture be separated?

7.4 DEMONSTRATION OF THE 'SALTING OUT' EFFECT

(emulsion breaking by addition of water soluble salts)

Required

Cedarwood oil; sodium chloride; sodium carbonate; propan-1-ol; 2-methylpropan-2-ol

Procedure

1 Add 5 cm^3 of cedarwood oil and an equal volume of distilled water to a small separating funnel and shake vigorously for one minute.

2 Allow the contents of the funnel to settle and observe the effect.

3 Add a little sodium chloride to the mixture, shake and again allow to stand for a few minutes.

4 Comment on your observations.

Sequel

1 Repeat the experiment using sodium carbonate instead of sodium chloride.

2 Try the effect of salting out solutions of propan-1-ol and 2-methylpropan-2-ol in water (use a 1:2 ratio of alcohol to water).

7.5 DETERMINATION OF THE OIL CONTENT OF PEANUTS

(Soxhlet extraction apparatus)

Required

Fresh peanuts; petroleum ether (b.p. range 40–60 °C)

Procedure

1 Weigh out sufficient crushed peanut kernels to half fill the Soxhlet thimble (fig 7.3).

2 Place the thimble in position and add sufficient petroleum ether to the boil-up flask to one-third fill it.

3 Heat the solvent using a hot plate or heating mantle until it is gently boiling and allow the extractor to run for an hour.

4 After allowing to cool, pour out the solvent from the lower flask into a weighed crystallizing dish, together with any extract left in the thimble compartment.

5 Allow the petroleum ether to evaporate off (**CARE:** extinguish all flames) and then reweigh the dish to find the yield of oil.

6 Calculate the percentage of oil present in the peanuts.

Sequel

Repeat the experiment using sunflower seeds and compare the results.

8 Chromatography

Chromatography (Gk. *chroma,* colour, *graphe,* a writing) describes processes which allow the resolution and identification of mixtures by the separation of their components into concentration zones. This is done by transporting the mixture across a stationary adsorbent using a moving liquid or gaseous phase. The mechanism of chromatographic separation lies in the repeated adsorption and desorption of the components of a mixture as it passes through a chromatographic system. This effectively multiplies any differences in the partitioning behaviour of the components and enables even complex mixtures to be resolved without difficulty.

The concept of theoretical plates has already been mentioned in connection with distillation (page 26) and may also be applied to chromatographic separations. Thus chromatographic systems can be thought of a comprising a number of theoretical plates each representing a particular equilibrium state between the stationary and moving phases. Because the number of theoretical plates possible in such systems is very large indeed the separation achieved is highly effective.

As the name suggests, this technique was originally confined to the separation of coloured substances, such as plant pigments or dyestuffs. More recently locating methods have been devised to enable the resolution of mixtures of colourless substances such as fats and aminoacids.

The first person to investigate the possibilities of chromatographic separation was the German dyemaster Runge. In his *Der Bildungstrieb der Stoffe* published in 1855, he produced artistic coloured patterns by the partial resolution of dye mixtures on filter paper. The first use of a chromatographic technique for analytical purposes is generally attributed to the Russian botanist Tswett, who in 1906 separated the components of a leaf pigment extract by passing it down a glass tube packed with powdered chalk. This process is now known as adsorption chromatography, because separation depends upon the differing affinities of an adsorbent for the components of a mixture. The value of this technique was not fully realized until the 1930s, however, when the pioneer work of Kuhn and Lederer in resolving the components of carotene attracted worldwide interest in this analytical tool.

In partition chromatography, introduced by the Nobel prizewinners Martin and Synge during 1941, separation depends upon the distribution of the components of a mixture between a moving solvent and water bound to a stationary phase such as silica gel. The application of partition chromatography was greatly extended by the discovery by Consden, Gordon, and Martin in 1944 that filter paper containing adsorbed moisture could be used to replace the original liquid-impregnated columns of adsorbent.

More recently gas partition chromatography has been developed in which partition of a gaseous mixture occurs between an inert carrier gas such as nitrogen and a column of non-adsorbing material moistened with a suitable liquid. Chromatographic separations can also be effected using ion exchange resin beds or by gel filtration, the latter involving the separation of species using molecular 'sieves' of varying pore size. In addition use has been made of electrical potential gradients to separate mixtures of ionic species on materials such as paper, cellulose ethanoate, and gels of starch or acrylamide, a process known as zone electrophoresis. Although not strictly a chromatographic technique, electrophoresis is included here because of its close association with chromatography.

Chromatography has proved such an elegant and effective technique, and is of such general application, that it is now used extensively in the fields of chemistry, biology and medicine.

Adsorption chromatography

In column chromatography the powdered adsorbent is packed into a vertical glass tube. A small quantity of the mixture to be resolved is added to the top of the column where it remains adsorbed. The chromatogram is then developed by allowing a solvent to flow down the tube. Separation of the components of the mixture takes place as a result of differential adsorption, and if the components are visible, coloured zones appear down the column. The column can now be extruded, the zones sliced off, and the separated components extracted from the slices with a suitable solvent.

More conveniently, the flow of solvent is continued until the components are eluted from the bottom of the column one by one. The different fractions are collected and the individual components recovered from the solvent (fig 8.1).

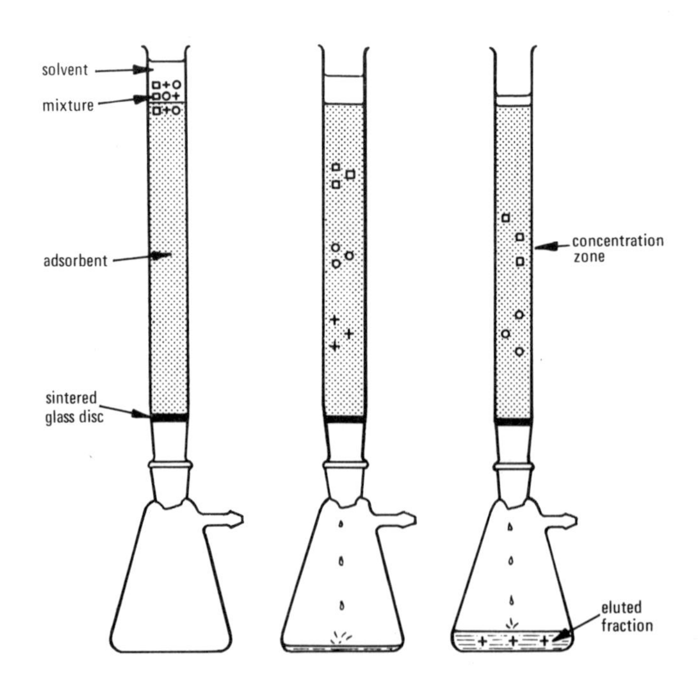

Fig 8.1 Resolution of a mixture by column chromatography
(i) mixture added to top of column
(ii) appearance of zones
(iii) collection of fractions

The comparatively large cross-sectional area of adsorbent used in column separations is particularly suitable for quantitative work.

A variant of column chromatography is commonly used in which a thin layer of the adsorbent acts as the stationary phase. The exploitation of this method is largely due to the pioneer work of Professor Stahl of Saarbrucken in 1958. A sheet of glass is coated with a thin layer of the adsorbent mixed with a binder such as calcium(II) sulphate. This is done by mixing the dry powder with a suitable liquid, usually water, to form a thick slurry, which is applied to the glass by spreading or dipping. The production of large numbers of standard chromatoplates for industry or research is carried out using a special applicator. After drying the plate, a little of the mixture is spotted just above one edge, which is then dipped into a shallow pool of solvent. The solvent is drawn up the adsorbent layer by capillary action, separation of the components of the mixture occurring as in a column chromatogram.

The popularity of thin layer chromatography (TLC) lies in the fact that it combines the convenience and simplicity of paper chromatography with the choice of a number of inorganic substrates normally used for column separation. Also TLC lends itself to the resolution of minute quantities of mixtures. The chromatograms are remarkably free from distortion as the fine state of division of the adsorbent prevents lateral diffusion. The quantities of adsorbent and solvent involved are very small, and the plates can be produced easily and quickly.

These advantages have made thin layer techniques attractive to the biochemist and forensic scientist. For example, minute traces of poison can be detected in a corpse, or the presence of additives in food confirmed. A recent application is the analysis of steroid metabolites in urine as the basis of an early pregnancy test.

Choice of solvent and adsorbent

The effectiveness of a chromatographic separation depends to a large extent upon the choice of suitable solvent. It may be possible to use the same solvent to dissolve the original mixture, develop the chromatograms, and elute the separated fractions. Usually it is necessary to use different solvents at each stage. As a general rule, non-polar* solvents (table D) such as petroleum are used for the original solution because they are readily adsorbed and substances are easily adsorbed from them. Polar solvents such as methanol resist adsorption, and are useful

Table D – Common solvents (in order of polarity)

non-polar ↓	1	petroleum ether
	2	tetrachloromethane
	3	cyclohexane
	4	ethoxyethane
increasing polarity	5	propane
	6	benzene
	7	organic esters
	8	trichloromethane
	9	alcohols
	10	water
	11	pyridine
highly polar	12	organic acids

*It should be noted that the term 'polarity' as used in chromatography does not have the usual chemical meaning. In this context polar solvents are those most readily moving along an adsorbent column and the polar components of a mixture are those most strongly adsorbed.

Table E – Some common adsorbents

Adsorbent	Strength	Used to separate
aluminium(III) oxide	strong	general use
magnesium(II) oxide	strong	general use
calcium(II) hydroxide	strong	carotenoids
silica gel	strong	sterols, fatty acids, glycerides, terpenes
charcoal	strong	sugars, peptides, aminoacids
chalk	medium	xanthophylls, carotenoids
calcium(II) phosphate(V)	medium	enzymes, proteins
sucrose	weak	chlorophylls, xanthophylls
starch	weak	enzymes
talc	weak	sterols, esters
cellulose	weak	sugars, dyes

for eluting the fractions from the column. Those solvent's with properties of an intermediate nature are used to develop the chromatogram.

It must be remembered, however, that this is not a hard and fast rule and that such factors as solvent power, volatility, and viscosity must be borne in mind when choosing a solvent. In addition, solvents behave differently on different adsorbents.

The adsorbent (table E) should preferably be colourless, cheap and readily available in a pure state. It must not be soluble in the solvents used, nor the components of the mixture, neither must it react chemically with them. The adsorbent power of the material is a function of its molecular structure, oxygen-containing groups being specially active in this respect. Thus sucrose is a weak adsorbent while aluminium(III) oxide has very powerful adsorbent properties. Usually an active adsorbent is most effective when used with a non-polar solvent and vice versa (fig 8.2). Often more effective resolution of a mixture can be obtained using a mixed column. In this case layers of two or more adsorbents are packed into the tube in succession.

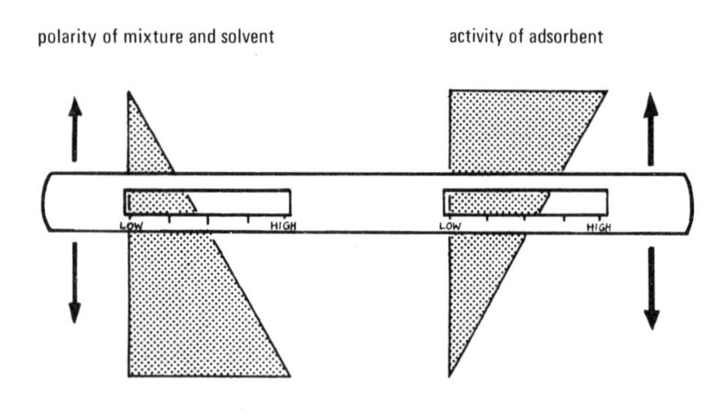

Fig 8.2 Visual guide to selection of solvent and adsorbent for a given mixture

Column chromatography

The simplest type of apparatus for column chromatography is a length of wide glass tubing, constricted at one end to pass through a cork in the neck of a Buchner flask. A wad of glass wool wedged in the constriction is used to retain the column of adsorbent, which half fills the tube. The upper half of the column is used as a solvent reservoir. Solvent flow may be increased by applying suction to the flask. This must be done with great care to avoid 'channelling' the column. It is more satisfactory to use pressure to speed up the flow of solvent. A simple device for this purpose is to fit a bicycle valve to a plug in the top of the separation tube. A hand pump can be used to maintain the required degree of pressure above the column.

More expensive, but more convenient for column chromatography, are glass ground joint assemblies for both macro and semi-micro separations (fig 8.3). These can be fitted with a sintered glass plate of porosity grade O to support the adsorbent. This becomes partially blocked during the packing of the column, allowing a suitable flow under either positive or reduced pressure.

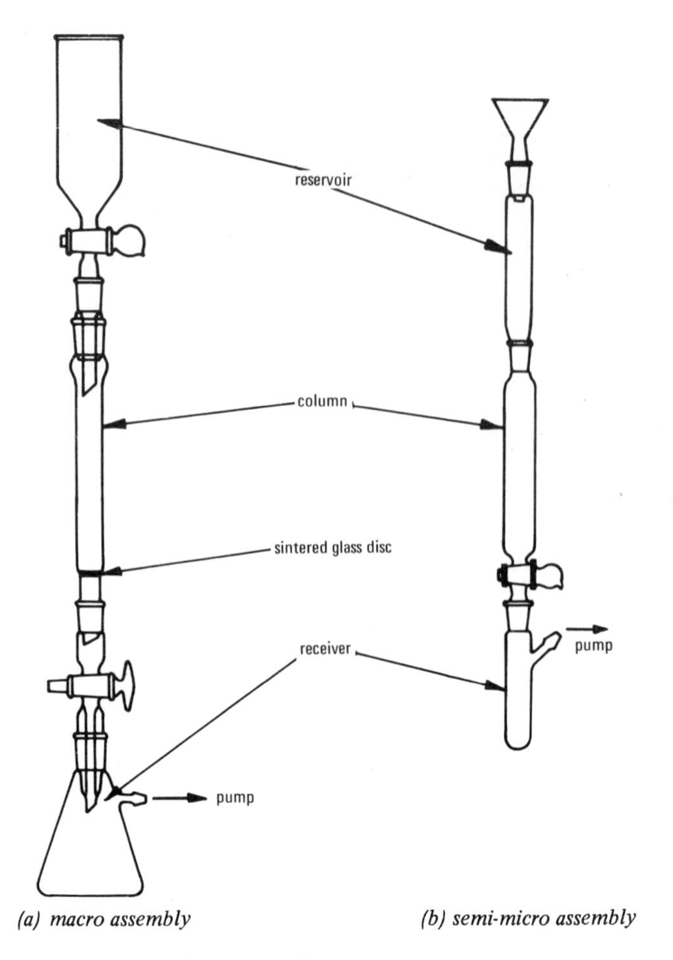

(a) macro assembly *(b) semi-micro assembly*

Fig 8.3 Column chromatography apparatus

A further advantage of ground joint apparatus is that a solvent reservoir can be fitted to the top of the column, thus enabling the whole of the latter to be packed with adsorbent. The semi-micro columns are usually about 1 cm diameter while for larger scale work columns of from 2 to 5 cm are used.

Micro columns can also be prepared using lengths of capillary tubing. In this case centrifuging is necessary to force the solvent through the column and offset the powerful surface tension effects (fig 8.4).

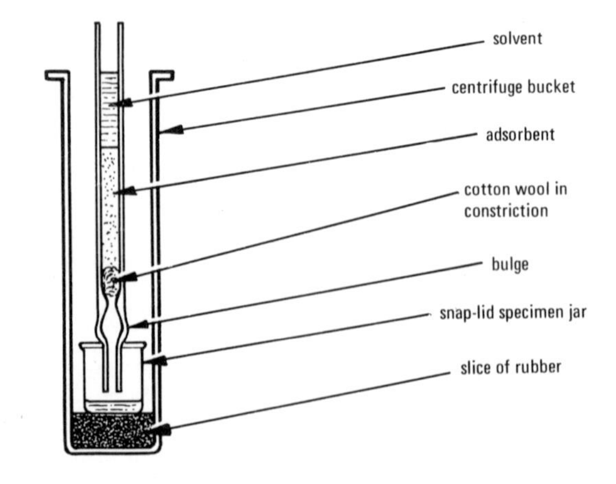

Fig 8.4 Micro-chromatographic tube

Packing the column with the adsorbent is an art which is only acquired with practice. A carelessly packed column inevitably gives a disappointing separation. There are two methods in general use.

Wet method (fig 8.5a). The adsorbent is poured into the solvent and stirred to form a smooth creamy suspension containing about one-third its volume of solvent. This is poured into the tube in small quantities. The solid settles out as a smooth, evenly packed layer as the solvent drains away. The process is repeated until the column reaches the desired height. If a mixed column is being prepared small circles of filter paper are dropped in to separate the adsorbent layers.

An alternative wet method is to half fill the tube with solvent and then pour in the adsorbent in a steady stream. In both methods it is most important not to let the surface of the column dry out before running the chromatogram.

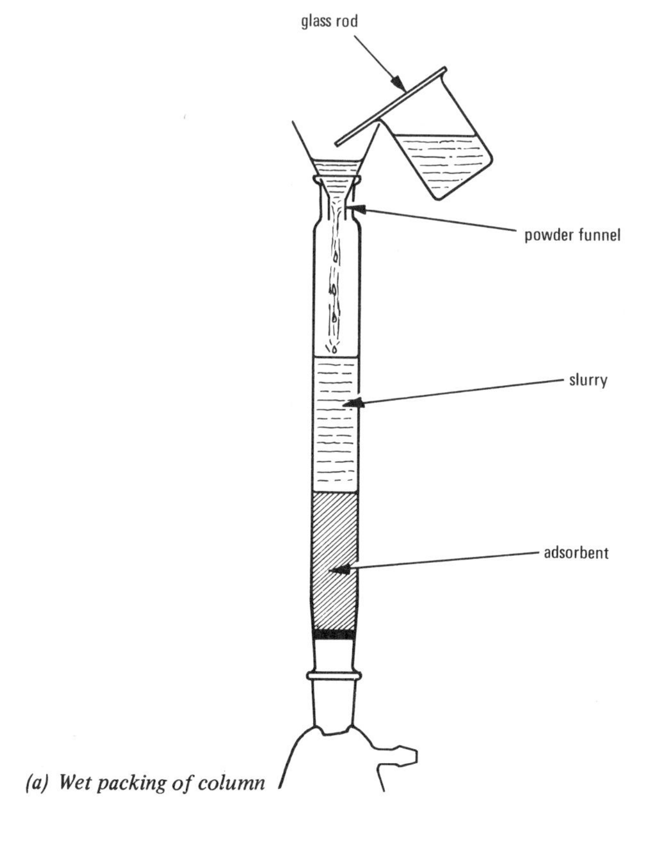

(a) Wet packing of column

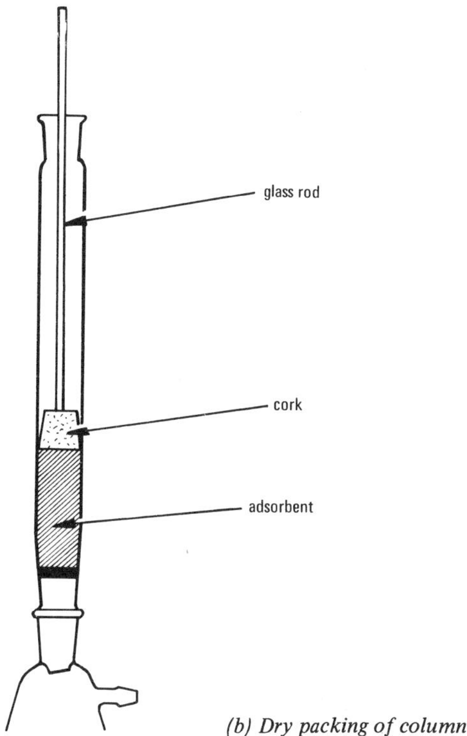

(b) Dry packing of column

Fig 8.5

Dry method (fig 8.5b). Small quantities of the adsorbent powder are poured into the column using a powder funnel. Care must be taken not to entrap tiny pockets of air during this process. After each addition the powder is tamped down using a small cork fixed to a length of dowel rod. The aim is to maintain an even packing of the column, and to leave the surface quite flat after each addition. The top of the column is then protected by a circle of filter paper and the solvent added. Once this has been done, the same precaution against drying out must be taken as before.

After the column has been prepared, a little of the mixture is pipetted onto the upper surface which should be just covered by solvent. The sample of mixture should be adsorbed as a shallow even layer, leaving the adsorbent undisturbed. Alternatively, a disc of filter paper can be impregnated with the mixture and dried, before application to the column surface. Interference with the solvent flow is prevented by perforating the paper disc with a number of pinholes before use.

The chromatogram is developed by running the solvent steadily through the column, using pressure or suction if this is necessary, and taking care to keep the column surface covered. If good resolution has been obtained, collecting the separated components of the mixture as they are eluted successively from the bottom of the column is a simple matter.

In industry or research when more complex mixtures have to be resolved, a device is used which automatically collects small successive fractions from the bottom of the column. These are retained in small tubes mounted on a revolving turntable.

Thin layer chromatography (TLC)

To obtain satisfactory results it is necessary to use adsorbents specially prepared for thin layer work. These may be inorganic substances such as silica gel, aluminium(III) oxide, or kieselguhr, but cellulose and a number of cellulose derivatives such as carboxymethyl cellulose and cellulose ethanoate are also suitable. An inert filler is often used to improve the quality of the film and to fix it to the glass plate. Fluorescent indicators or other special additives such as silver(I) nitrate(V) are occasionally added to the adsorbent material to facilitate the identification or resolution of the mixture components. Before use the powder is mixed to a thick slurry in a flask or mechanical mixer using distilled water or chloroform and then loaded into a spreader. The mixture must be used immediately because it sets in about four to five minutes.

The standard size 20 cm square plates are held in a frame and covered with an even layer of the slurry using a single steady movement of the spreader. About five plates at a time are treated and they are then air dried or gently heated. Heating is necessary in any case when the mixture to be separated contains hydrophobic ('water-hating') components, a temperature of about 120 °C being maintained for 20 to 30 minutes. This 'activates' the plates which can be kept in a desiccator until required. The handling of the chromatoplates is greatly simplified by the use of special racks fitted with carrying handles in which the wet plates can be stored.

The plates are marked before use by drawing a line 2 cm

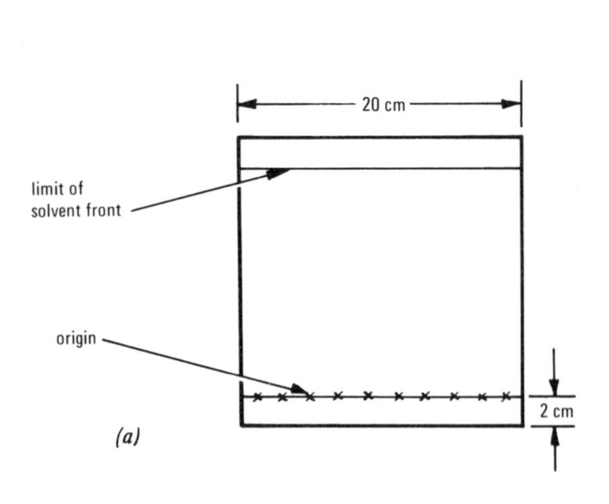

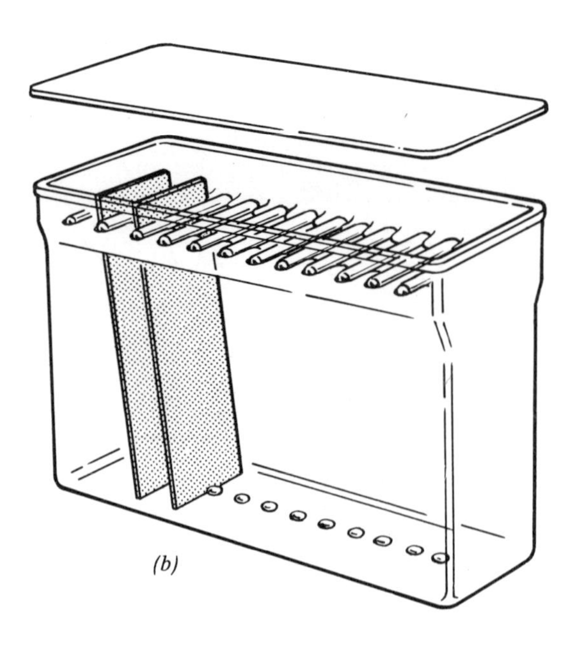

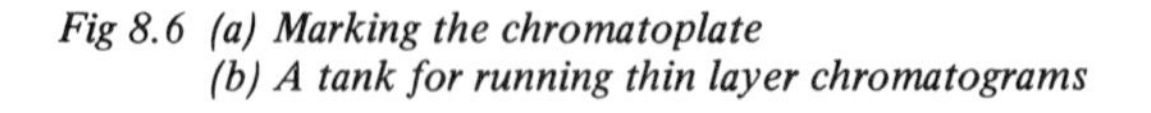

Fig 8.6 (a) Marking the chromatoplate
(b) A tank for running thin layer chromatograms

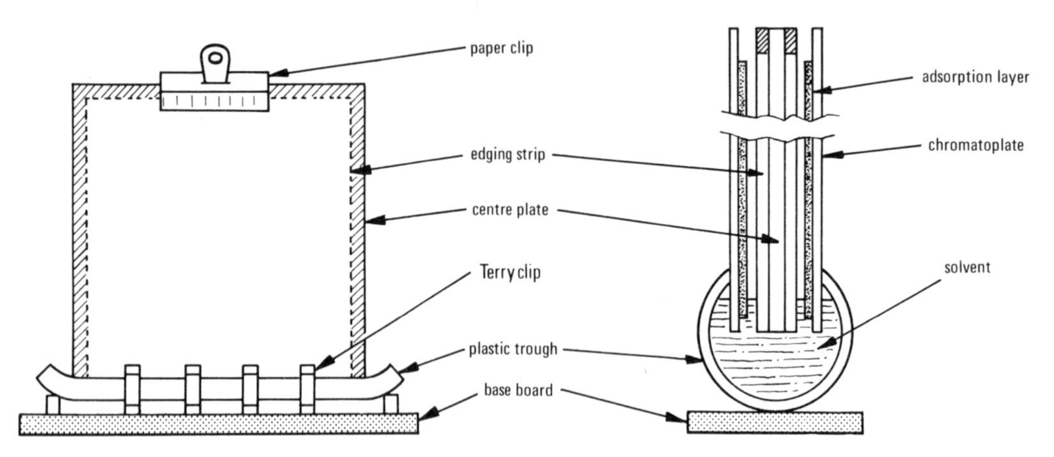

Fig 8.7 Double S-tank for TLC

from one edge to indicate the limit of the solvent front, and samples of the mixture spotted in a line 2 cm from the opposite edge and a similar distance apart. A transparent plastic template makes marking and spotting easier and more accurate, and short lengths of melting-point tube or small wire loops are also helpful in dispensing uniform samples of mixture on the plates.

The spots must be kept as compact as possible and concentrated, if necessary, by adding further drops of the mixture, care being taken to dry the spots thoroughly between each addition. A hair-dryer is useful for this purpose. For the separation of larger amounts of material the mixture is spread along the origin as a line of small dots (fig 8.6).

The chromatoplates are developed by standing them on edge in a pool of solvent on the floor of a glass tank. The atmosphere in the tank is kept saturated with vapour by lining the walls with a few strips of glass fibre paper or filter paper soaked in solvent. The solvent rapidly travels up the plates which are removed and dried as soon as the marked line is reached. The final position of the solvent front on the chromatoplate is thus marked by the limit line. If the distance of the solvent front from the origin is taken as unity the distance travelled by each separated component can be expressed as a decimal fraction known as its R_f value. Since this is constant for a given solvent under identical experimental conditions, it affords a valuable means of identification.

A simple and inexpensive 'double S-tank' for the simultaneous development of a pair of thin layer chromatography plates has been described by Bancroft*. This consists of a 20 x 20 cm glass sheet having 1 cm wide strips of glass cemented (with Araldite) around the upper and side edges on both sides. Chromatoplates can be laid on either side of this glass sheet and held in position by a large paper clip, the prepared layer fitting into the cavity formed by the edging strips. The solvent trough can be made from a 30 cm length of 18 mm bore PVC tubing sealed at either end, and held firmly on a blockboard base by four Terry clips. A cork borer is used to make two holes in the plastic trough 20 cm apart, and the holes joined by a straight slit. After adding the solvent to the trough the plates are clipped to the double S-tank and wedged into the slit in the trough. The apparatus has been used for two-dimensional development (fig 8.7).

Commercial spreaders are expensive but it is quite easy to make one. An effective construction has been described by Collings†. This consists of a 2 x 24 cm x 1 m blockboard base fitted on the long sides with 2.5 x 1 cm plastic foam strips. The glass plates rest on the foam rubber and are held in position while spreading by clamping down with two strips of 4 cm Handy Angle. The spreader is a simple Perspex box 18.5 x 7.5 x 7.5 cm with

Chem and Ind,* 1966, **15, 621.
†*Chem and Ind,* 1966, **14,** 576.

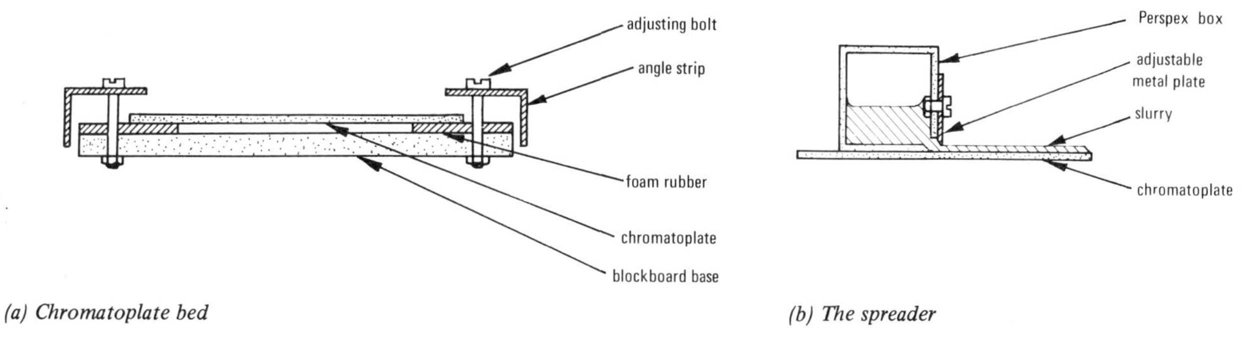

(a) Chromatoplate bed *(b) The spreader*

Fig 8.8

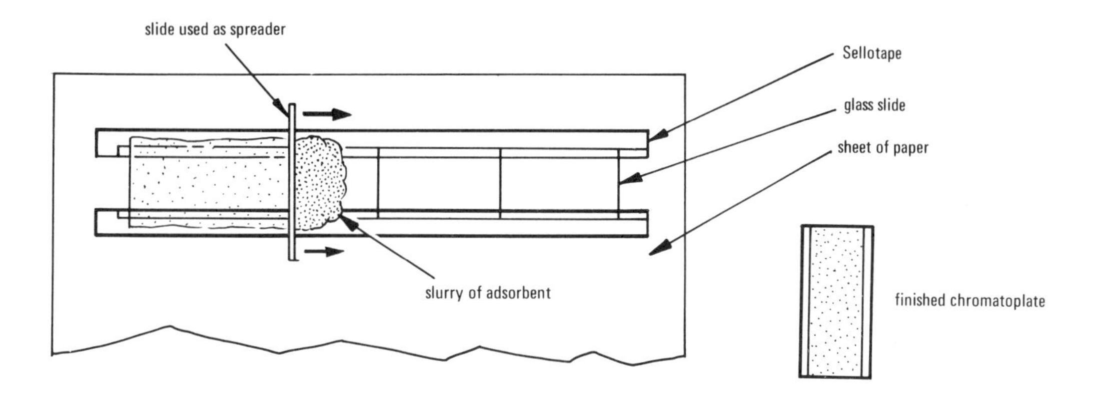

Fig 8.9

an adjustable metal plate on the trailing edge to enable the thickness of the layers to be varied (fig 8.8).

An even simpler technique for preparing chromatoplates for school use is to use microscope slides Sellotaped lengthwise along their edges to a piece of paper. The slurry is poured on one end of the line of slides and is spread evenly and in one movement, using the longer edge of another slide (fig 8.9). The thickness of the film can be regulated by using additional layers of Sellotape along the glass edges. When dry, the tape is stripped off leaving the chromatoplates with a clean edge.

Thin sheets of plastic or glass fibre coated with a layer of adsorbent and ready for use can be purchased (Kodak Eastman) but these are comparatively expensive. Glass tubes coated on the inside with a layer of adsorbent are also available.

Loev and Goodman* have outlined an interesting 'dry column' chromatographic technique in which separations comparable to that obtained by TLC can be achieved quickly using either the traditional glass column or a section of nylon tubing. Thin plastics tubing eliminates many of the difficulties associated with the use of glassware and is considerably cheaper. The nylon film is transparent to ultraviolet light, is strong and inert, and can be cut to any length and heat sealed.

The mixture to be resolved is added to the top of the dry column and allowed to soak in before developing with a solvent in the usual way. Development is complete when the solvent reaches the bottom of the column. The nylon tube can then be cut longitudinally and the column sliced up. This method has been used successfully with alumina columns of up to 1.8 m.

Trial methods

It is useful to be able to run a number of rapid small scale chromatograms using different solvent and adsorbent pairs. In this way the combination giving the most effective resolution can be determined before carrying out a large scale separation. A simple technique is to pour a little of the adsorbent against the side of a Petri dish, forming a small sloping wedge. The mixture is then spotted on the centre of the thin edge of the wedge and allowed to dry. A little of the trial solvent can then be run into the dish and its resolving power noted (fig 8.10a).

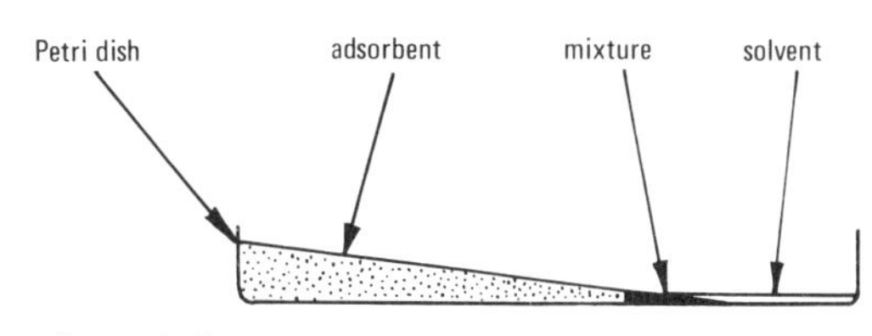

(a) wedge technique

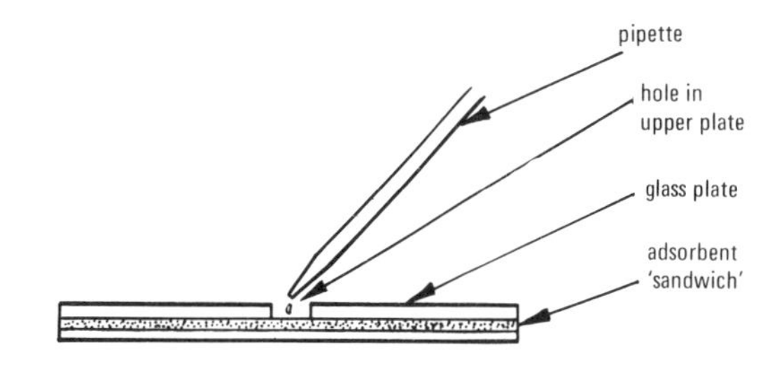

(b) plate technique

Fig 8.10 Trial methods

An alternative method is to sandwich a small layer of the adsorbent between two sheets of glass about 14 cm square. A little of the mixture is then dripped through a small hole drilled in the centre of the upper plate and allowed to dry. The trial chromatogram is then developed by adding solvent through the hole using a small pipette (fig 8.10b).

Chem and Ind,* 1967, **48, 2026.

Paper chromatography

The resolution of a mixture by partition chromatography relies upon differences in distribution of the components between two immiscible solvents. One of these solvents is invariably water which is held stationary in the pores of a supporting medium such as cellulose. The second solvent moves across the adsorbed water and a continuous partition of the components of the mixture takes place between the two liquids. Those substances which are more soluble in the moving solvent tend to be carried on while those having an affinity for water are held back. Because the redistribution of solutes between the two solvents takes place many thousands of times as the solvent moves forward, small differences in the distribution coefficients of the components of the mixture produce an appreciable degree of separation.

As in adsorption chromatography, the different affinities of the components of a mixture for the supporting medium also produce a separating effect, but this is less important than the distribution effect because a weakly adsorbent material is always used.

A great advance in the use of partition techniques for separating mixtures chromatographically was the introduction of the so-called 'open column'. This made use of filter paper sheets as the supporting medium for the stationary water phase — even 'dry' paper containing up to 12% water bound to the cellulose fibres. The technique has a number of advantages. Only a small volume of the moving solvent is required and this moves across the paper by capillarity or gravitational force, suction or pressure being unnecessary. Also the paper is strong enough to be self supporting, and regular enough in thickness to prevent the distortion of the bands which often occurs when using column separations. The method is also simple and cheap and can be carried out with the minimum of apparatus. In addition to this, several chromatograms can be run on a single sheet, and locating agents (see page 45) can easily be applied by dipping or spraying.

A small drop of mixture is spotted near the lower edge of a strip of filter paper which is then dried. The end of the strip is dipped into a shallow pool of solvent, care being taken not to immerse the spot of mixture. In ascending chromatography the solvent is now allowed to rise up the strip by capillary action. In descending chromatography the paper is hung from a small trough of solvent, so that movement of the liquid is aided by gravity. Both methods of separation are carried out in a closed glass container, or cabinet, to avoid loss of solvent by evaporation.

As the mobile solvent flows across the spot of mixture, it carries the components along the paper strip. The distance that each substance moves depends upon its solubility in the moving solvent, its degree of adsorption by the paper, and its distribution between the mobile solvent and the water bound to the cellulose fibres of the paper. As with thin layer chromatography (fig 8.12) the R_f values of the separated components are useful for identification purposes.

Ascending chromatography

Cleanliness is essential for success in paper chromatography. The laboratory bench should be covered with a sheet of clean paper to avoid contamination with grease and dirt. Also the chromatographic paper should be handled as little as possible and then only by the edge furthest away from the origin.

Before the mixture is applied to the paper strip, the origin of the chromatogram must be marked. This is done by drawing a pencil line lightly across the paper about 3 cm from the bottom edge. A small pencilled cross is then marked on the centre of the line. If a sheet of paper is to be used for a multiple separation then a number of points are marked at 3 cm intervals.

The mixture is best applied to the origin marks using a small length of nichrome wire twisted at one end into a loop. Before use, the loop is cleaned by rinsing in distilled water and then heating in a bunsen flame. A little of the liquid is picked up in the loop by dipping it into the mixture and this is then spotted on the paper. It is desirable to keep the spot as compact as possible. If further applications are needed to concentrate the spot of mixture it must be carefully dried before each additional drop is added.

There are a number of ways in which the marked strip can be suspended while the chromatogram is developed. Most convenient to use are the pieces of equipment specially designed for the purpose, but it is very easy to improvise. A gas jar makes an excellent container for either a paper cylinder or strips. Short lengths of glass rod fitted with rubber tubing 'stubs' can be used to suspend paper strips, or alternatively a glass hook held in a cork, or two sheets of glass. Fig 8.11 illustrates some of these alternatives.

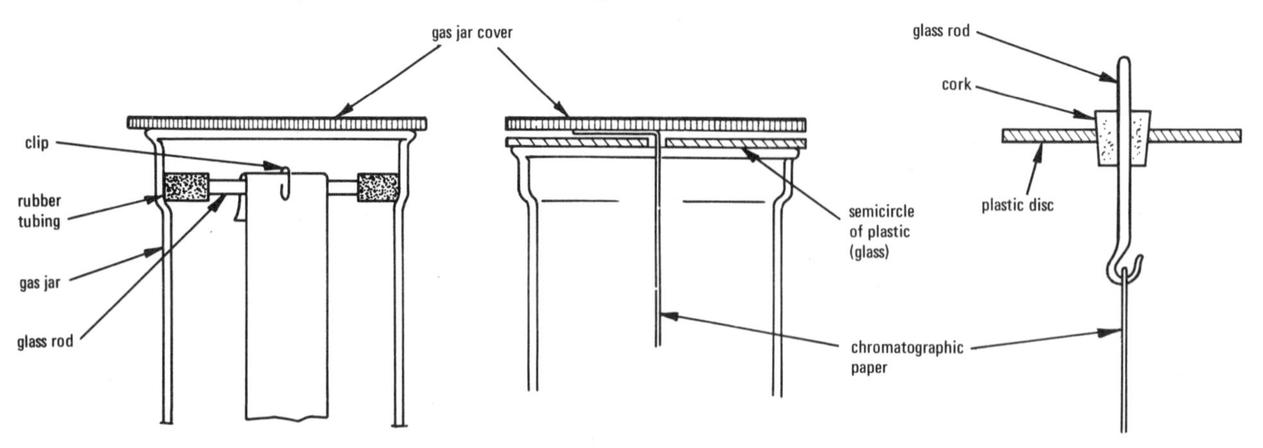

Fig 8.11 Alternative methods of suspending chromatograph paper

The marked edge of the paper is now immersed in the solvent, taking care to keep the origin above the level of the liquid. The strip must also be kept clear of the sides of the container. To reduce evaporation of solvent from the surface of the chromatogram while running, it is necessary to saturate the atmosphere of the container before use. This can be done by adding the solvent to be used to the covered empty container and allowing it to stand for a few hours.

The solvent rises quite quickly and separation begins at once. After running the solvent for about an hour, or when the solvent front has reached the desired point on the strip, it is necessary to fix the position of the separated components by drying. This may conveniently be done by hanging the strip in a fume cupboard with the extractor fan working. The drying process can be speeded up by using a small hair-dryer. Before drying is complete the solvent front must be marked with a pencil line so that the R_f values of the separated components can be calculated.

Normally a chromatogram is developed until the solvent front is a convenient distance from the origin. This is not always possible and sometimes the origin–solvent front distance is an awkward value thus making the calculation of R_f values tedious. It is helpful in this case to use a piece of rubber band marked in cm from 0–10 with white ink. R_f values can be readily determined in this way by laying the band alongside the chromatogram and stretching until the zero and 10 cm marks align with the origin and solvent front respectively.

The details of the mixture separated and the solvent used should also be pencilled on the dry chromatogram, which can then be trimmed if necessary and mounted using adhesive transparent tape. It is also wise to pencil round the separated components to record their R_f values in case of fading (fig 8.12).

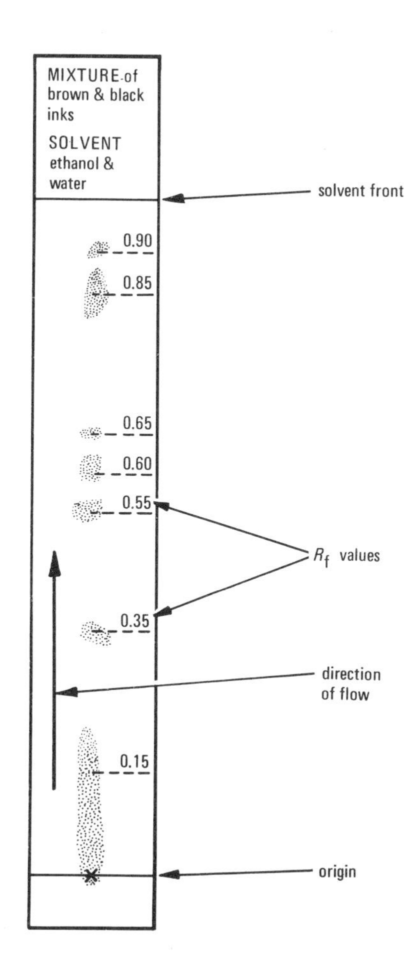

Fig 8.12 Ascending chromatography

A number of ascending chromatograms can be run simultaneously using a special sheet of paper which is divided into strips by a number of narrow vertical slots. After spotting, the sheet is rolled into a cylinder and stood on end in the solvent which is contained in a jar (fig 8.13).

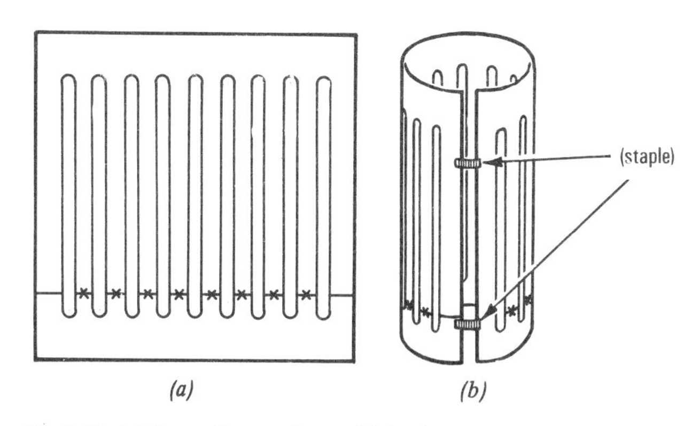

Fig 8.13 (a) Paper for running multiple chromatogram
(b) Paper folded ready to start the solvent run

Descending chromatography

For descending chromatography, a trough of solvent is used in which the end of the paper strip or sheet carrying the spots of mixture is anchored by a short length of glass rod. The paper hangs downwards over another glass rod which holds it away from the side of the trough and prevents siphoning of the solvent. The solvent rises up the strip by capillarity and after passing over the anti-siphoning rod is carried across the spot of mixture and down the strip by a combination of capillarity and gravity (fig 8.14).

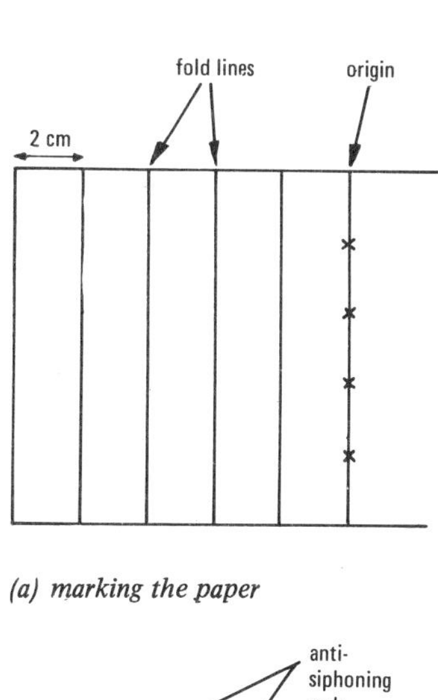

(a) marking the paper

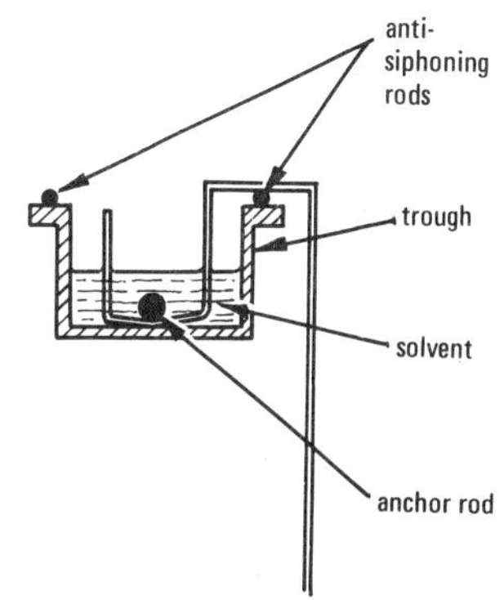

(b) paper positioned in the trough

Fig 8.14 Descending chromatography

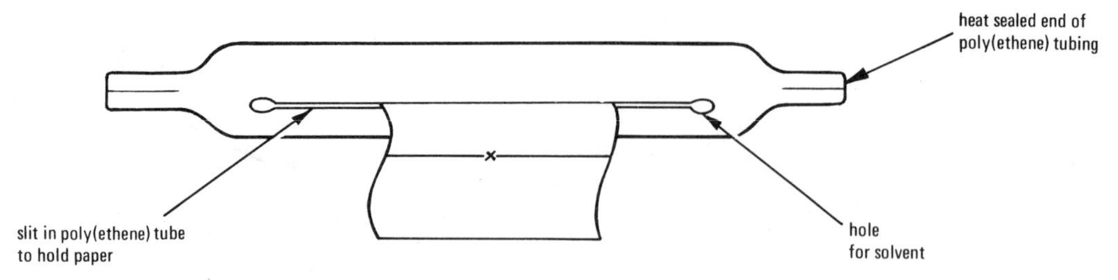

Fig 8.15 Simple poly(ethene)trough for descending chromatograph

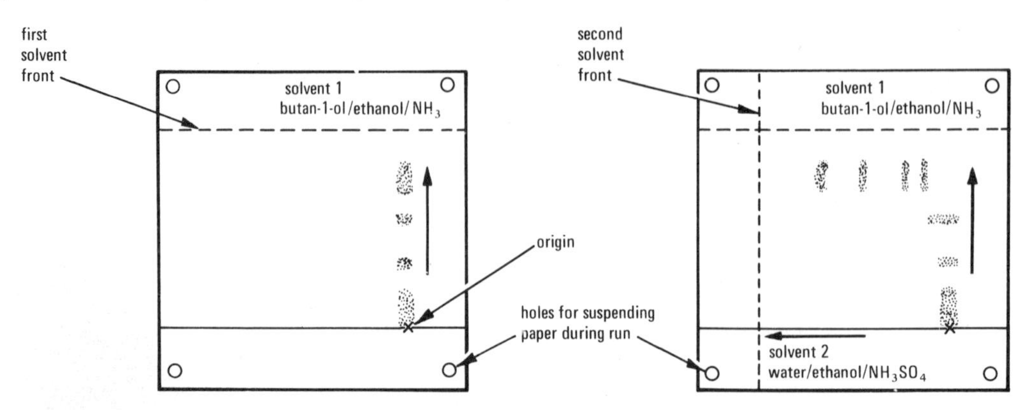

Fig 8.16 (a) Chromatogram after running the first solvent (b) Chromatogram after running second solvent at right angles

When marking papers for descending chromatography it is helpful to draw five pencil lines at intervals of about 2 cm, with the origins marked on the fifth line. This is to leave room for the end of the paper to be tucked into the trough.

As with ascending separations it is convenient to use specially made apparatus for running descending chromatograms, but with a little ingenuity satisfactory results can be obtained using home-made equipment. One simple method of making a solvent trough is to seal the ends of a large bore section of poly(ethene) tubing with a soldering iron and then cut a slit along the top for insertion of the paper (fig 8.15). A hole can be bored at either end of the slit with a cork borer, which not only prevents the slit elongating but facilitates the addition of solvent.

If it is required to recover components after separation by paper chromatography, this is best done by cutting the appropriate zones from the dry chromatogram after running, and extracting each component in a Soxhlet, or by simply rinsing in a suitable solvent. If the resolved mixture is colourless, a thin strip can be cut off one side of the chromatogram and the positions of the required components revealed by a locating technique (page 46).

Two-dimensional chromatography

It may happen that the R_f values of some of the components of a particular mixture are almost identical for a given solvent. This results in poor resolution of the mixture due to incomplete separation. To overcome this difficulty an extension of the paper strip technique is used, known as two-dimensional or two-way chromatography. This is achieved by running two different solvents in succession at right angles to each other.

The mixture is spotted close to one corner of a square of chromatographic paper as shown in fig 8.16, and the first solvent run to produce an ordinary chromatogram. The paper is then dried, and turned through 90° so that the second solvent travels at right angles across the path of the first chromatogram, thus producing a further separation.

The technique has a high resolving power because not only is the chromatographic path longer, but the R_f values of the components of the mixture vary according to the solvent used. If the two solvents are carefully chosen, substances resolved poorly during the first separation will be satisfactorily separated during the second run.

Metal frames can be obtained which enable several separations to be carried out simultaneously, the sheets being dried and turned without being removed from the frame. More simply the sheets can be threaded on glass rods spaced with stubs of poly(ethene) tubing and hung in a glass tank or shallow sink. Alternatively they can be rolled into tubes and the separations carried out using a gas jar as a container. Fig 8.17 illustrates this.

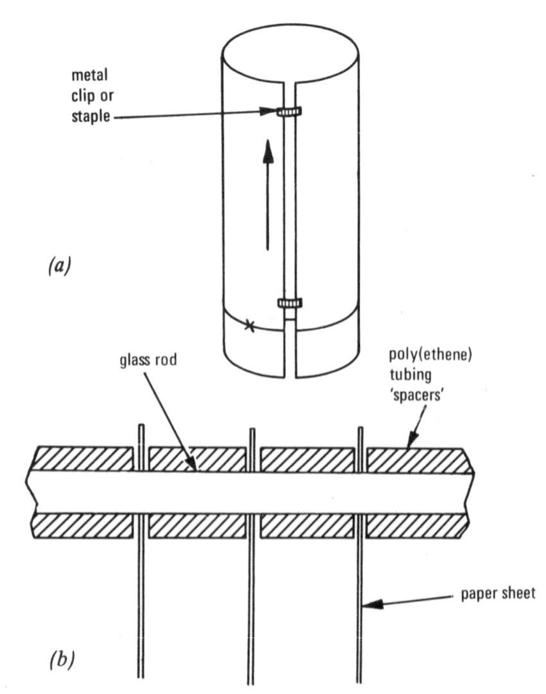

Fig 8.17 (a) Paper rolled into a cylinder for running one- or two-way chromatograms in a gas jar
(b) Method of suspending sheet chromatograms

After drying, the two-way chromatograms should have details of the solvents used written alongside arrows leading from the origin and showing the direction of solvent flow (fig 8.16).

Paper disc chromatography
Chromatographic separations can be quickly and simply carried out using paper discs instead of strips. Ordinary filter paper circles are quite satisfactory for this. The mixture is spotted at the centre of the disc, and the solvent fed to this point by means of a paper wick. The separation zones appear as a series of concentric rings around the origin. An advantage of this method is that the best sector can be chosen by inspection, and cut from the final chromatogram for identification purposes or recording. Three techniques are commonly used.

1 A filter paper circle has a narrow sector removed except for a short stub near the centre which is left as the wick. This is bent at right angles to the plane of the paper. The mixture is then spotted at the centre of the circle and dried. The paper is laid across the lower half of a Petri dish containing the solvent. The upper half of the dish is now carefully laid over the paper to prevent loss of solvent by evaporation. The wick should just dip into the solvent, the paper being removed and dried before the solvent front reaches the outer edge of the dish (fig 8.18a).

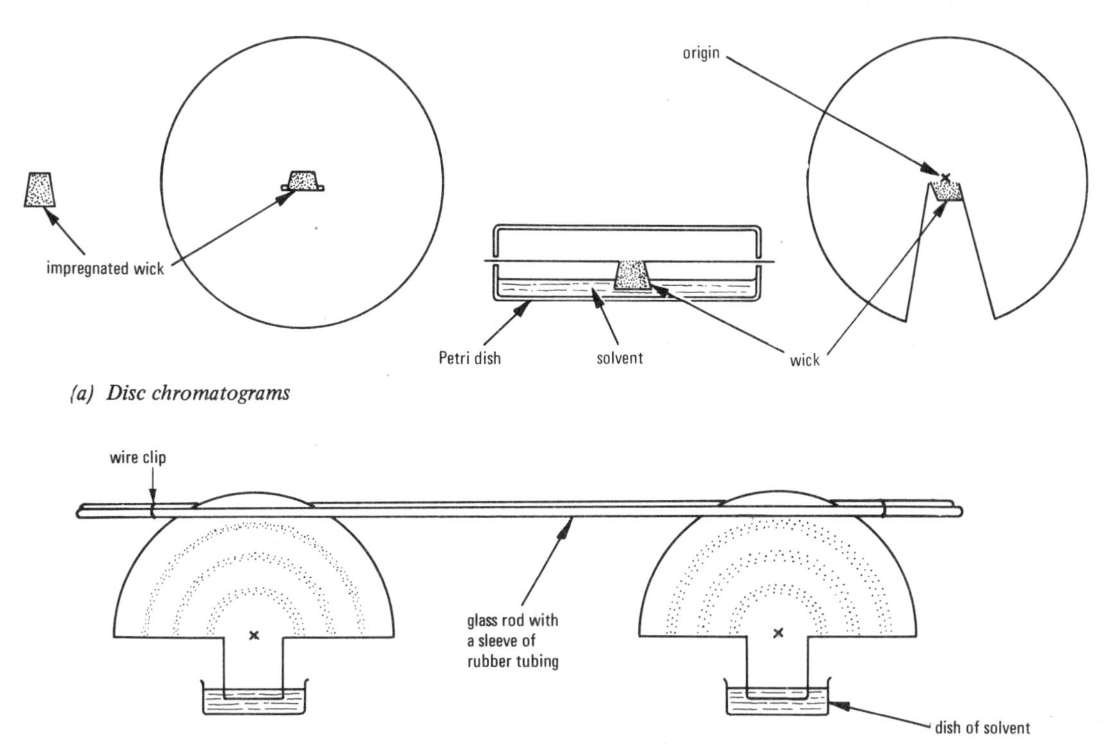

(a) Disc chromatograms

(b) Alternative technique for running disc chromatograms

Fig 8.18

2 An alternative to the above method is to cut a small wedge of filter paper, soak this in the mixture and then dry it. By pushing the impregnated wick through a slit cut in the centre of a paper disc, a chromatogram can be run in the same way as described in **1.**

3 In this method no container is used to prevent evaporation and the wick is therefore made much wider to increase the flow of solvent. Mushroom-shaped sections are cut from paper discs and these are suspended stem downwards by nipping their top edges between two glass rods sheathed in rubber tubing. The mixture is spotted at the top of the stem which dips into a small dish of solvent (fig 8.18b).

An interesting type of paper disc chromatography which allows the rapid analysis of up to five samples simultaneously was first demonstrated in Cambridge in 1956. In this technique, which is due to Kawerau, specially cut filter discs are divided into five equal sectors by radial slots. The samples are spotted on the origins of the sectors and the chromatograms run in a special container, the solvent being supplied to the centre of the disc by a capillary glass tube which also acts as a support.

Reverse-phase chromatography
Mixtures containing components which are insoluble or only sparingly soluble in water cannot be resolved by the usual chromatographic techniques. This is because adsorption does not take place on the stationary phase and the components are carried straight through the system by the solvent front. To overcome this problem a special technique termed reverse-phase chromatography is used. This involves treating chromatographic paper before use with silicone oil, rubber latex, or paraffin to provide a stationary adsorbent phase for the water insoluble components. The eluting solvent used in reverse-phase chromatography is usually an organic liquid containing a small amount of water.

Location of colourless substances

The term chromatography is misleading because the process can be applied equally well to the resolution of mixtures containing colourless components. In fact mixtures of colourless substances are much more likely to be encountered. After resolving colourless mixtures, however, it is necessary to treat the chromatogram in some way to render the separated components visible.

Some substances which are colourless or only faintly coloured in daylight fluoresce strongly under ultraviolet light. Other substances absorb ultraviolet light and appear as dark spots or bands. When using the thin layer technique, fluorescein may be added to the adsorbent powder before use. The spots on the chromatoplate then show as dark areas when viewed under ultraviolet light, owing to their quenching effect on the fluorescence. Plates prepared with fluorescein in this way can also be used to detect the presence of unsaturated compounds by treating the dried plate with bromine vapour. Fluorescence under ultraviolet light only occurs at the site of the unsaturated material when the bromine has been absorbed.

With paper chromatograms, permanent records may be made by pinning them over a sheet of photographic paper and exposing to an ultraviolet source for a few seconds. The paper can then be developed in the normal way. X-ray film has been used similarly to record chromatograms of radioactive substances. Thin layer chromatograms can be stripped off the glass support plate using a sheet of transparent Fablon. This can then be mounted on paper to give a permanent record. An alternative method of recording chromatograms is to make use of a Xerox copier or similar device.

The advantage of these physical methods of location is that they do not damage the separated components. Chemical methods on the other hand rely upon the formation of coloured derivatives with specific reagents. They are, however, rapid, of wide application and require little special apparatus.

With column chromatograms, the adsorbent can be extruded and streaked along its length with a locating reagent, which can afterwards be scraped off if the fractions are to be eluted. Thus vitamin A can be detected by streaking with a saturated solution of antimony(III) chloride in trichloromethane. Another common technique is to run an indicator solution through the column before carrying out the separation. This is useful for locating substances such as fatty acids.

Locating reagents may be applied to paper chromatograms either by dipping or spraying. Thus aminoacids are revealed as pink or blue spots by spraying with a dilute solution of ninhydrin in propanone, and heating. Similarly, sugars can be detected using alkaline silver(I) nitrate(V) solution. Care has to be taken not to disturb or remove the separated components.

If inorganic adsorbents are used in thin layer separations, the presence of organic substances such as lipids can be revealed by heating with sulphuric(VI) acid, which produces charring at the site of the spots of separated materials. A record of the intensity and location of these spots can be obtained using a photodensitometer. This is a photoelectric scanning device which translates the chromatogram into graphical form, enabling a more accurate comparison of results than is possible by eye.

Ion exchange chromatography

Certain insoluble materials contain labile ions which can be exchanged with certain of the ions present in solutions passed over them. This process is called ion exchange and was first appreciated by soil scientists such as Way and Lemberg during the nineteenth century. Commercial use of ion exchange was first made by Gans in 1905, who used the naturally occurring mineral zeolite (greensand) to soften water. An important aspect of this process was that when 'exhausted' the zeolite could be regenerated by treatment with sodium chloride solution.

By the 1920s, chemists were attempting to produce synthetic ion exchange materials to replace the rather unstable zeolite type minerals. The first substance of this kind, produced by treating coal with sulphuric(VI) acid, was described by the German Liebknecht in 1934. In the following year two British chemists, Adams and Holmes, developed a number of novel synthetic ion exchange materials by condensing polyhydric phenols or phenolsulphonic acid with methanal. Originally these substances could only be used for cation exchange, but later it was shown that by condensing polyamines and methanal, anion exchange resins could also be prepared. Although the original anion exchangers were prepared by Adams and Holmes from aromatic amines such as benzene-1,4-diamine it was later found that higher capacity resins could be produced by incorporating aliphatic amines into the polymer network.

In 1944, D'Alelio showed that stable ion exchange resins could be based on a cross-linked poly(phenylethene) framework. Such resins were remarkably stable chemically and resistant to wear and tear. For example by treating beads of poly(phenylethene) with concentrated sulphuric(VI) acid a sulphonated polymer (Zerolit 225) is produced, which is an acidic cation exchange resin.

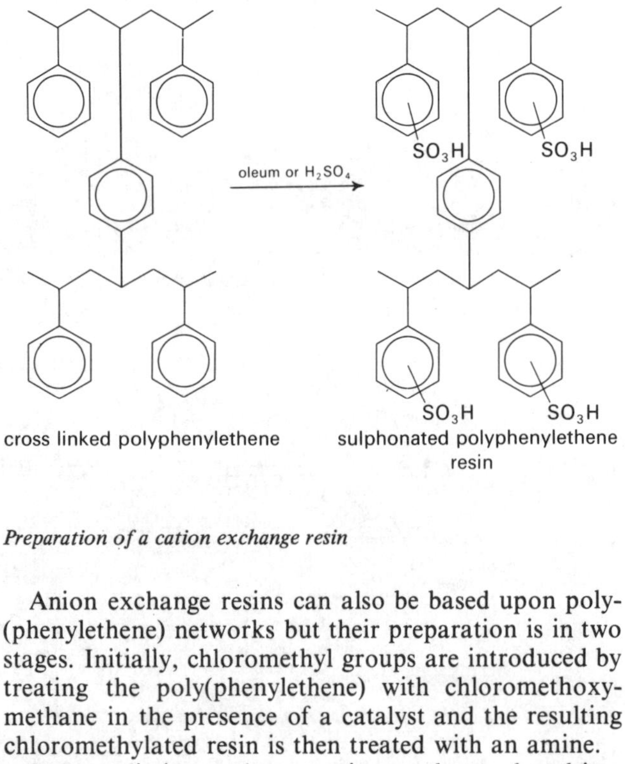

oleum or H_2SO_4

SO_3H SO_3H SO_3H SO_3H

cross linked polyphenylethene

sulphonated polyphenylethene resin

Preparation of a cation exchange resin

Anion exchange resins can also be based upon poly(phenylethene) networks but their preparation is in two stages. Initially, chloromethyl groups are introduced by treating the poly(phenylethene) with chloromethoxymethane in the presence of a catalyst and the resulting chloromethylated resin is then treated with an amine.

Polyacrylic ion exchange resins can be produced in a similar way.

More recently ion exchange resins have been produced in sheet form (Permaplex) and these have a number of important applications such as the desalting of electrolyte solutions. Mixtures of both anionic and cationic exchangers are also available (Bio-deminrolit) which enable water to be completely demineralized. Another recent innovation has been the introduction of isoporous resins in which a porous cross-linked structure is produced by a post-polymerization reaction within the resin bead. This has the advantage of being resistant to organic 'poisoning'.

CH_3OCH_2Cl / $AlCl_3$ catalyst

cross linked polyphenylethene

$N(CH_3)_3$ / tertiary amine

CH_2Cl CH_2Cl

chloromethyl derivative

$Cl(CH_3)_3NCH_2$ $CH_2N(CH_3)_3Cl$

anion exchange resin

Preparation of an anion exchange resin

Both the acidic (cation exchange) and basic (anion exchange) resins can be subdivided into those which ionize strongly over the greater part of the pH range, and the weak acid and weak base resins which only ionize fully above and below pH 7 respectively and are therefore restricted in their use.

The labile ions of an ion exchange resin are held in the polymer network by electrostatic forces of attraction. They can be exchanged with other ions of the same sign. Thus an acidic cationic resin can have its negatively charged sulphonic acid groups associated with hydrogen ions (hydrogen form) or sodium ions (sodium form). Similarly an ionic resin can be in the free base form when the basic groups are associated with hydroxyl ions, or in the chloride form after treatment with sodium chloride solution. Manufacturers specify the form in which resins are marketed (usually the sodium form and chloride form) and it is a simple matter to convert the resin to an alternative form before use if this is necessary.

The 'exchange capacity' of a resin is dependent upon the number of ionizable sites on the resin framework, the total capacity being expressed as milligram equivalents per unit of dry weight or wet volume — the latter being used when the resin is swollen ready for use. The term 'operating capacity' is also used to indicate the total capacity of a particular ion exchange unit during one cycle of operation.

Ion exchange resins can be used to resolve mixtures of ionizable substances by chromatographic techniques and are particularly suitable for the separation of amino-acids, sugars, proteins, and nucleotides. Industrially ion exchange is widely used for the softening and demineralization of water, the treatment of toxic effluent from factories, and for the recovery of valuable metals such as gold and uranium. An interesting application for this latter technique was the separation of rare earth elements for the Manhattan project* in World War II. Medical applications of ion exchange include the use of 'artificial kidney' devices to remove metabolic waste from the blood of patients with diseased kidneys.

A recent development in the field of ion exchange chromatography has been the introduction of ion exchange celluloses in sheet form. These enable resolution of mixtures to be carried out by paper chromatographic techniques. Also a special type of chromatographic paper has been produced which has a coating of ion exchange material.

Experimental details

Ion exchange resins are usually supplied in the form of moist beads and must be prevented from drying out by careful storage in a tightly closed jar or poly(ethene) bag.

*The code name for the development of the first atomic bomb.

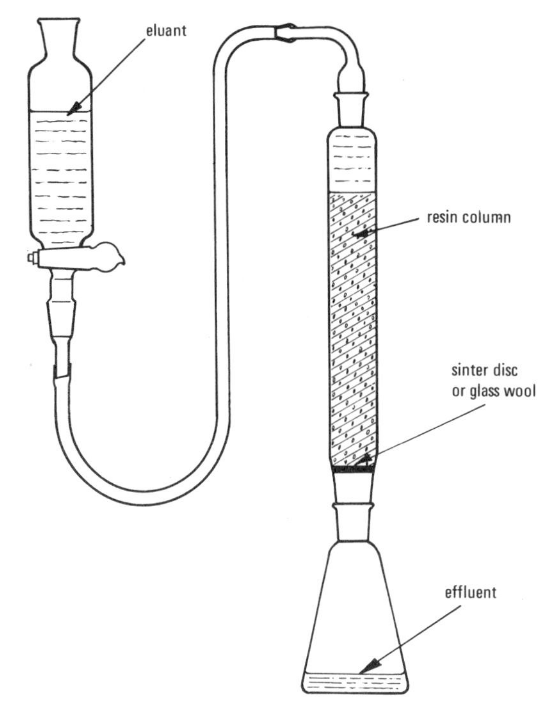

(a) Single column ion exchange apparatus

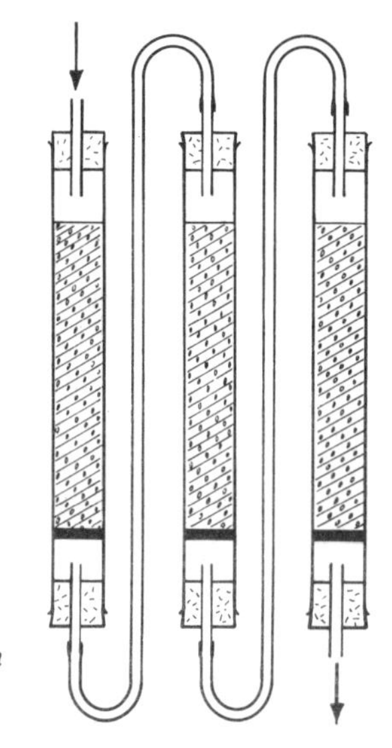

(b) Multiple column

Fig 8.19

Before use the resin must be washed. This is done by stirring the beads into a beakerful of distilled water and allowing to settle. The supernatant liquor containing any small fragments and impurities is sucked off using a tube attached to a filter pump. Washing is repeated until the wash liquor is clear.

The next step is to pack the column ready for the separation. Usually this can best be done by pouring a slurry of resin and water into a chromatograph tube and allowing the resin beads to sediment out. When packing tall narrow columns care must be taken not to trap air bubbles in the resin as this upsets the even flow of liquid and leads to 'channelling'. Sometimes 'back washing' is used to produce a satisfactory packing by forcing water up through the column to fluidize the resin bed.

Before use the column should be depleted and regenerated to remove undesirable impurities. With strong acid and strong base resins this can be achieved by respectively washing with 10% solutions of hydrochloric acid and sodium hydroxide, followed by rinsing with distilled water. Weak acid or weak base resin may be similarly treated with more dilute solutions of about 3% concentration before use.

As with column chromatography, the surface of the resin column must not be allowed to dry out or be disturbed when running in the mixture and this can be prevented by resting a disc of filter paper or a thin plastics disc on the top of the resin bed. Special care must be taken with anion exchange resins because they are only slightly denser than water and therefore easily disturbed. Sometimes multiple columns are used to give a sharper separation, the effluent from one column being run into the next (fig 8.19).

Gas chromatography

Gas chromatography represents an extension of chromatographic techniques to include the separation of gaseous mixtures, a carrier gas such as nitrogen or hydrogen being used as the moving phase instead of a liquid.

Early chromatographic separations of gaseous mixtures made use of gas–solid techniques in which the components of the mixtures were differentially adsorbed on the surface of a porous solid such as activated charcoal. A mixture of hydrocarbons was successfully separated by Claesson in this way as early as 1946. In the next few years gas–solid chromatography was developed by Phillips and others, and proved to be an analytical tool of surprising versatility and accuracy. In 1952, however, a paper was published by Martin and James outlining a gas–liquid chromatographic process which rapidly gained popularity and eclipsed the earlier gas–solid technique.

The column used for gas–liquid chromatography is a narrow bore looped tube packed with small crumbs of a porous solid such as kieselguhr (Celite) or crushed firebrick (Stermachol). The packing is coated with a thin film of a non-volatile liquid as the stationary phase. As in paper chromatography, separation is due to differences in the distribution of the components of the mixture between the moving phase (in this case a gas) and the stationary liquid phase. In gas–liquid separations the solid merely acts as a support for the liquid phase.

The temperature of the column can be varied from below freezing point to as high as 400 °C according to the nature of the sample to be analysed and the liquid stationary phase being used. The sample is introduced at one end of the column, usually by means of a hypodermic needle inserted through a self-sealing rubber cap, and eluted by the carrier gas. As in other chromatographic processes the components of the mixture are successively eluted from the end of the tube. Because the separated fractions are gaseous and present in such minute quantities, special detecting devices are required which usually depend upon changes in the density, thermal conductivity, or degree of ionization of the gas stream.

A further development in the field of gas chromatography was made by Golay, who in 1957 proposed the use of capillary tubes coated with a thin film of the stationary phase. The use of capillary columns of this type proved very successful and enabled a very high degree of resolution to be obtained.

The efficiency of gas chromatographic columns can be expressed in terms of theoretical plates in a manner analogous to that already discussed in connection with distillation columns (page 26). Here the height of a theoretical unit (HTU) is determined by dividing the length of the column by the number of theoretical plates.

Within the space of two decades gas chromatography has become the most useful of all the chromatographic techniques and is now extensively used for the analysis of minute amounts of volatile substances or gaseous mixtures. Samples of essential oils of the order of 10^{-7} g have been successfully resolved in this way, and polymers identified by analysis of the gaseous products of destructive distillation.

Large scale gas chromatography is also being used for preparative purposes, and continuous separation techniques have been developed in which a moving liquid phase is used.

Experimental details

In gas–liquid chromatography (GLC) the tube is usually of glass or stainless steel, although aluminium and copper are also used. The filling of crushed firebrick or diatomaceous earth is first washed to remove dust and other unwanted material ('fines') and then thoroughly dried in an oven. The stationary liquid phase is dissolved in a volatile solvent such as ethoxyethane and the required amount of dry solid stirred in. The proportion of liquid phase to solid phase is not critical, a ratio of 1:4 by mass being normally used. After thorough mixing the volatile solvent is allowed to evaporate off and the treated solid is gently heated in an oven to a few degrees above the intended operating temperature. This should be about 50 °C above the boiling point of the least volatile component to be separated.

After tapping the filling down into the column the latter can be coiled and is usually operated at a fixed temperature. Although liquid baths boiling under reflux were originally used, electrical heating systems are now common. More sophisticated stainless steel columns are sometimes heated by passing an electric current directly through them.

Detectors

There are several types of detector in use, each relying on some physical property of the issuing gas. Flame ionization and thermal conducting detectors are most commonly used. These are known as differential detectors because they register a difference in response between pure carrier gas and that containing eluted components.

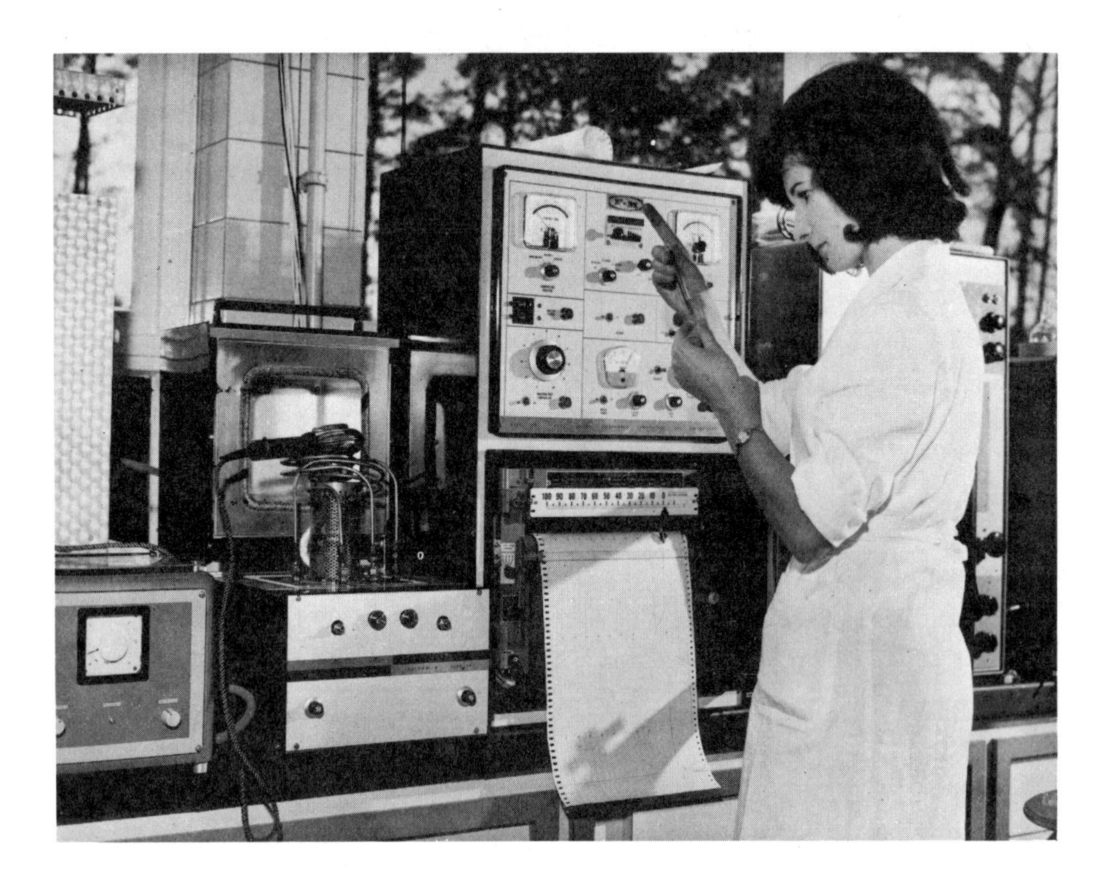

Gas chromatography equipment

Flame ionization detector. This is the most commonly used type of detector (fig 8.20). Most organic substances, even those which are not flammable, ionize in a flame of hydrogen gas and therefore increase its electrical conductivity. By continuously monitoring the conductivity of the flame, the appearance of separated components can thus be detected. A potential difference of about 150–200 V is maintained across the flame, the metal jet acting as the negative electrode and a platinum wire mounted near the tip of the flame as the positive.

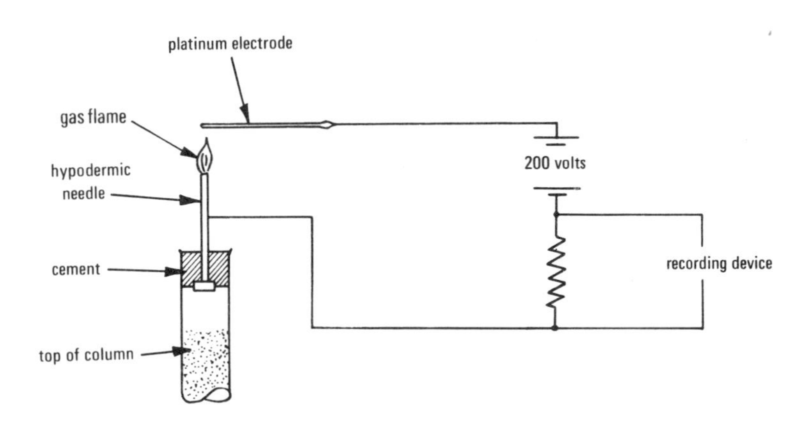

Fig 8.20 Simple type of flame ionization detector

Thermal conductivity detector (katharometer). Metals such as platinum show considerable changes in resistance with relatively small alterations of temperature. The principle of the katharometer is based upon this relationship between temperature and electrical resistance. A short length of platinum wire heated by a low voltage current of electricity is fixed in the gas stream issuing from the column. Any alteration in the composition of the effluent gas also changes its thermal conductivity. This alters the temperature of the platinum and hence its resistance. These changes in resistance are continuously monitored using a bridge type circuit, the signal from which is recorded by a galvanometer or pen recorder, thus indicating the appearance of eluted components in the carrier gas.

Columns

Capillary columns. Capillary columns for GLC are usually made from nylon, glass, stainless steel, or copper, and may be as long as 120 m. The stationary liquid phase is dissolved in a volatile solvent such as ethoxyethane to give a 10% solution. A little of this solution is drawn into the capillary until about 1% of the tube has been filled. The solution is then blown through the tube by a current of dry gas. The solvent evaporates to leave a very thin layer of the stationary phase on the capillary walls. The running of the chromatogram and detection of components is carried out in the same manner as with the packed column. but separation is more complete and can

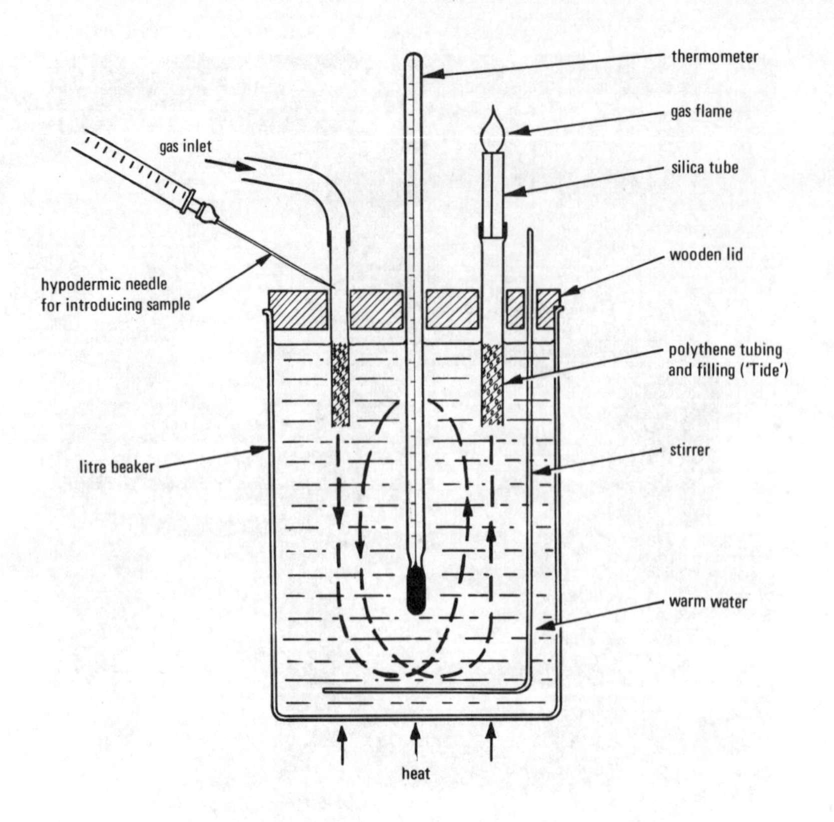

Fig 8.21 Simple apparatus for gas chromatography

be performed at lower temperatures and at greater speed. Capillary columns with up to one million theoretical plates have been reported.

'Tide' columns. A very simple piece of apparatus to demonstrate the principles of GLC can be constructed using household detergent powder as the filling for the column (fig 8.21). Powder of this type normally contains up to 20% water which acts as the stationary liquid phase. The detergent is first sieved through a piece of wire gauze to remove any dust and then packed into a length of poly-(ethene) tubing. The ends of the tube are passed through two holes in a wooden lid which fits on the top of a litre beaker, and also has holes for a thermometer and stirrer. The coiled tube should fit snugly inside the beaker without touching the walls or bottom. One end of the tube is then attached to a gas tap and the other fitted with a short length of silica tube as a jet (a piece from a silica triangle is suitable). Before use the beaker is filled with water and heated to about 50–60 °C. The gas is then turned on and the jet ignited to give a small non-luminous flame.

A mixture of low boiling point hydrocarbons can be injected with a hypodermic needle as shown, the arrival of the separated components at the jet being indicated by a yellowing and increase in size of the flame.

Chromatographic gels

Another interesting development in chromatography has been the introduction of gels obtained by copolymerizing dextran with varying amounts of epichlorhydrin (Sephadex). The gel can be imagined as a loose irregular network formed from chains of glucose residues cross-linked by glyceryl bridges derived from the epichlorhydrin molecules. Such a structure is uncharged at neutral pH values because the hydroxyl units on the dextran chains do not ionize under these conditions. At neutrality therefore the gel acts as a micro filtering unit or 'molecular sieve', but under acid conditions other factors such as adsorption and ion-exclusion come into play. The proportion of epichlorhydrin determines the degree of cross-linking produced on the polymer gel and this in turn affects the size of the molecular sieve produced and the water adsorption of the resin (water regain). Thus Sephadex G 25 has a small pore size which will exclude molecules having a relative molecular mass larger than about 4000. Sephadex G 200 on the other hand can accommodate molecules with a relative molecular mass of up to 200000, and has a water regain of just over 20 times its dry mass, compared to that of 2.5 for Sephadex G 25.

Chromatography using columns of Sephadex gels involves either 'crude' separations, where large molecules unable to enter the molecular sieve are isolated from

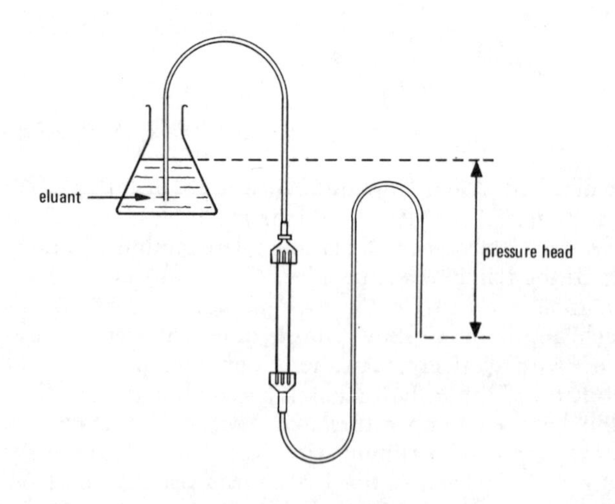

Fig 8.22 Experimental set-up for gel filtration

smaller molecules able to enter the gel pores, or 'refined' separations when the molecules are all small enough to enter the gel but pass down the column at different speeds due to variations in their rate of diffusion.

The technique has been successfully used in the separation of polysaccharides, aminoacids, nucleotides, and peptides.

Zone electrophoresis

When an electric current is passed through a solution of an ionizing substance, the charged ions migrate towards the electrode with the opposite charge. If the current is turned off, orderly migration ceases and the ions diffuse in all directions. In the process known as zone electrophoresis the ionizing particles are 'frozen' in zones when the current is stopped and can therefore be separated from one another. This is achieved by streaking or spotting the ionizing material across the centre of a piece of chromatographic paper soaked in a conducting buffer solution and bridging two electrode compartments which are also filled with buffer solution. Across the ends of the conducting strip an electrical potential is maintained which may be as high as 10 000 V d.c. (10 kV) in high voltage electrophoresis. Evaporation of the buffer, which is accelerated by the heat generated by the current passing along the strip, is reduced by either sandwiching the strip between two slightly greased glass plates or by carrying out the separation in a closed container.

After passing the current for some time the non-ionizing substances remain close to the point of origin, although some movement may take place due to diffusion and the flow of water towards the cathode (electro-osmotic effect). The charged ions, however, will move away from the origin towards the electrode of opposite charge. The rate of migration of the ions is determined mainly by their size and charge, small highly charged ions moving the fastest. This results in the ions being concentrated into bands or streaks. When a satisfactory separation has been achieved the current is turned off and the paper dried, thus preventing the ions from re-mixing by diffusion. The strip can now be cut up and the fractions removed by solvent extraction if required. Locating agents have to be used in the case of colourless components. In the case of proteinaceous mixtures, dyestuffs such as Amidoblack are often used for this purpose since they are strongly mordanted by the protein components which are revealed as coloured zones on rinsing the treated electrophoretogram.

The first record of electrophoresis being used to resolve mixtures chromatographically was in 1937 when the biologist Konig separated the proteins in snake venom by this means. The use of filter paper as a supporting medium was well established by the late 1940s and since then other supporting media such as agar, starch and polyacrylamide gels, and cellulose ethanoate have been used. Low voltage electrophoresis has been widely used for the resolution of complex mixtures containing serum and plasma proteins or aminoacids.

Two-way chromatograms can be run using a combination of electrophoresis and paper chromatography. Thus an initial separation can be produced by electrophoresis and a second run made at right angles using a solvent. Conversely a paper chromatogram can be further separated by electrophoresis at right angles to the original solvent run.

Continuous electrochromatography can be carried out when it is required to separate larger quantities of material. A sheet of chromatographic paper with a number of V-shaped notches cut from the lower edge is hung by the upper edge from a trough of buffer solution. The buffer travels down the paper and an electric current is made to pass across the solvent path by dipping the opposite corners of the sheet (curtain) into two electrode troughs. The mixture is applied continuously to the top of the sheet using a micro-pipette, or a motor-driven hypodermic syringe. As the components of the mixture are carried down by the flow of solvent, they are also moved across the paper in varying degrees by the electrical field. Thus by the time they reach the bottom of the sheet a lateral separation of the components of the mixture has taken place and these can be collected separately as they drip from the notches on the lower edge of the paper curtain (fig 8.23).

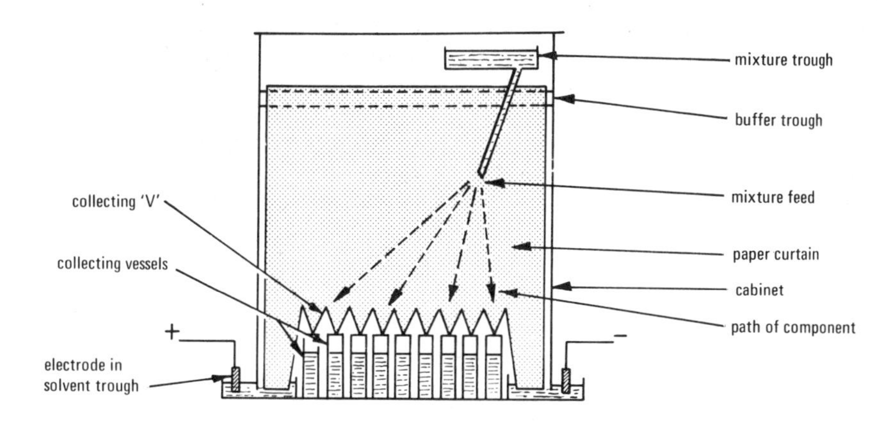

Fig 8.23 Apparatus for continuous electrophoretic separation

Experimental details

The apparatus available for low voltage electrophoresis is usually based on the vertical and horizontal designs shown in fig 8.24. The strip of filter paper to be used has an origin marked across its centre with a pencil and a cross marked on the end to be placed in the anode chamber. Care must be taken at this stage only to handle the strip by the ends. The paper is then passed through a small beaker containing the buffer solution to be used, drained for a few moments to remove the excess liquid and then blotted between clean dry blotting paper. The saturated paper strip is carefully laid across the supports so that the origin is centrally situated and the ends are in

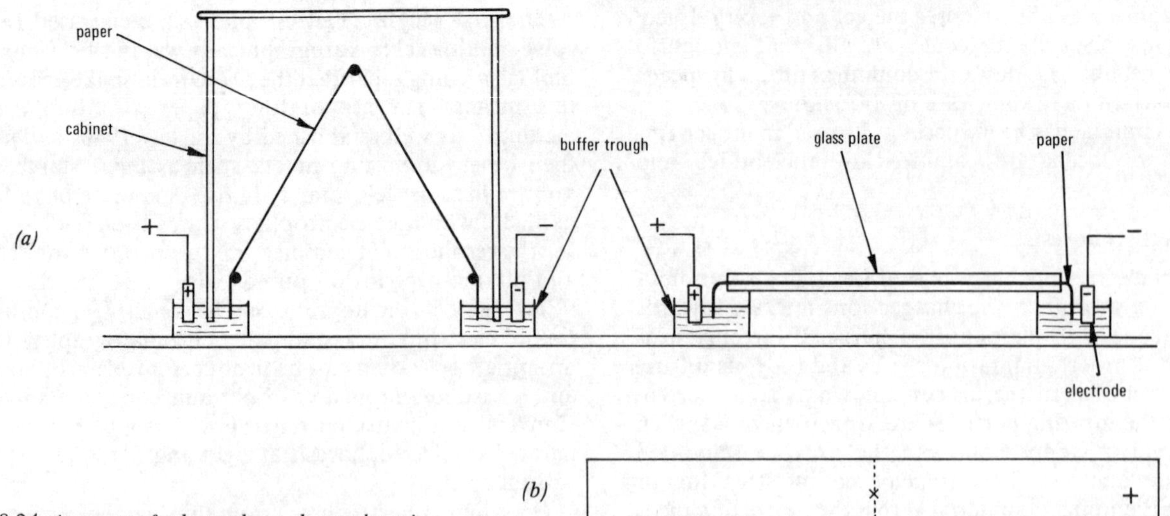

Fig 8.24 Apparatus for low-voltage electrophoresis

the buffer compartments, the end marked + being in the compartment connected to the anode.

In the best types of apparatus the electrode compartments are only indirectly in contact with the buffer solution by means of a wick bridge or a piece of synthetic sponge. This limits diffusion of the buffer electrolysis products along the paper and also helps to keep the ends of the paper at constant pH. The paper in the immediate vicinity of the origin is next lightly blotted and the mixture applied either by streaking or blobbing. A micropipette or the sealed end of a melting-point tube are convenient for this purpose.

If it is necessary for the mixture to be applied more than once, the strip must be dried before each application. In this case the buffer is not applied until after the final application. This is best carried out by dipping the strip into the buffer liquid leaving a narrow width on either side of the origin. The dry area then becomes wetted by capillary action from either side, and the streak of mixture is not disturbed.

The current should be switched on as soon as possible after preparing the strip, care being taken not to touch the apparatus while the current is on. The d.c. power supply is best obtained from a unit specially designed for the purpose. After running for a suitable length of time the current is switched off and the paper blotted and then rapidly dried in an oven or with an infrared lamp. The separated substances can then be located on the resulting electrophoretogram.

A very convenient apparatus for carrying out horizontal electrophoresis on a number of samples simultaneously has been designed by Dr Kohn of Queen Mary's Hospital, Roehampton. This comprises a water-jacketed Perspex tank fitted with a sloping lid to avoid drops of condensation falling on the strips during running. The base of the tank is fitted with four compartments for the two electrodes and the two buffer reservoirs, adjacent buffer and electrode compartments being connected electrically by means of a wick.

The paper (or cellulose ethanoate) strips are held firmly at their ends by shoulder pieces, sagging of the strip being prevented by three sets of adjustable plastics pins. A current of about 0.4 μA per cm width of strip at a potential of 200 V is recommended for serum proteins, separations of up to 7 cm being achieved in about two hours, depending upon buffer strength and temperature.

Chromatography experiments

8.1 SEPARATION OF METHYL ORANGE AND METHYLENE BLUE AND A MIXTURE OF 2- AND 4-NITROPHENYLAMINE

(semi-micro column, wet packing)

Required

Methyl orange; methylene blue; ethanol (industrial methylated spirit); aluminium(III) oxide (chromatographic grade); 2- and 4-nitrophenylamine

Procedure

1 Dissolve 0.5 mg of methyl orange and 2 mg of methylene blue in 1 cm^3 of ethanol.

2 Push a small tuft of glass wool into the column of the semi-micro assembly shown in fig 8.3 and then wet pack with a slurry of aluminium(III) oxide in ethanol (fig 8.5).

3 Fill the tap funnel with ethanol. Allow the solvent to drain from the column until it is within 1 mm of the top of the adsorbent and then introduce the dye mixtures.

4 When the dye has been almost adsorbed, run in the solvent slowly from the reservoir.

5 Allow the solvent to run until all the methylene blue has been eluted from the column, using suction if necessary.

6 Empty the ethanol from the solvent reservoir and replace with water.

7 Remove the receiver flask with the eluted methylene blue and replace with an empty one.

8 Pour water down the column until the methyl orange has also be eluted.

Sequel

1 Evaporate the two eluted solutions to dryness in tared evaporating basins using a water bath heated by an electric hot plate. Find the mass of recovered dye and record the percentage efficiency of recovery.

2 Wet pack a second column using a slurry of aluminium(III) oxide in benzene and separate a mixture of 0.2 mg each of 2-nitrophenylamine and 4-nitrophenylamine in 1.5 cm^3 of benzene. Continue eluting with benzene until the two compounds, visible as yellow bands, have been successively washed from the bottom of the tube.

3 Recover both eluted fractions and concentrate by distilling off most of the benzene*. Pour the concentrated solutions into a watch glass and evaporate to dryness in a fume cupboard over a beaker of boiling water heated by an electric hot plate.

4 Identify the two compounds by their melting points (2-nitrophenylamine m.p. 72 °C; 4-nitrophenylamine m.p. 148 °C).

***CARE: toxic and flammable vapour.**

8.2 SEPARATION OF LEAF PIGMENTS

(mixed column, dry packing)

Required

Tomato leaves (or spinach); petroleum ether (40–60 °C and 60–80 °C b.p. range); methanol; aluminium(III) oxide (chromatographic grade); calcium(II) carbonate (precipitated chalk); sucrose (castor sugar)

Procedure

1 Finely chop a few tomato leaves (or spinach) and extract the leaf pigments by steeping overnight in a mixed solvent containing 100 cm^3 of petroleum ether (60–80 °C b.p. range) and 35 cm^3 of methanol.

2 Use the macro column assembly shown in fig 8.3.

3 Fill the lower third of the column with aluminium(III) oxide, tapping the tube gently on the bench and finally ramming down with a cork ramrod (fig 8.5b).

4 In a similar way fill the middle and upper parts of the tube with calcium(II) carbonate and sucrose respectively.

5 Filter the leaf mash using a Buchner funnel and shake the extract with three 50 cm^3 portions of water to remove the methanol.

6 Add a few granules of anhydrous calcium(II) chloride and shake in a stoppered flask to dry.

7 Filter the dried extract and concentrate to a volume of about 1 cm^3 by evaporation at low pressure. To do this heat the solution over a water bath in a stoppered filter flask connected to a filter pump by its side arm.

8 Next fill the reservoir of the column assembly with petroleum ether (40–60 °C b.p. range) and saturate the column, applying suction by means of a filter pump attached to the lower flask. When a flow rate of about 20 drops a minute has been obtained remove the reservoir, allow the solvent level to fall to within 1 mm of the column surface and add the concentrated leaf extract.

9 Replace the reservoir and develop the chromatogram with the petroleum ether until the pink band of the carotenes has reached the lower region of the column.

10 Stop the solvent flow and suck the column dry with the pump.

Sequel

1 See if you can recognize the olive green β-chlorophyll in the upper part of the column, the lighter green α-chlorophyll in the middle section, the yellow xanthophylls in the lower middle, and the bottom band of pink carotenes. The layers can be scratched out of the tube one by one and eluted with a 2% solution of methanol in ethoxyethane if time permits.

2 Examine the column under ultraviolet light if possible and comment on the effect produced.

8.3 ANALYSIS OF BLACK AND BROWN INKS

(thin layer chromatography)

Required

Black Quink ink; brown ink; ethanol (industrial methylated spirit); aluminium(III) oxide (TLC grade)

Procedure

1 Fill a 250 cm^3 beaker to a depth of 0.5 cm with a 1:4 mixture of ethanol and water, cover with a watch glass, and allow to stand for a few minutes before use.

2 Make a slurry of aluminium(III) oxide with water and prepare three chromatoplates from microscope slides using either a spreader (fig 8.8) or the simple technique shown in fig 8.9.

3 Dry the chromatoplates in a warm oven, and gently mark the origin with a soft pencil about 1 cm from the bottom edge.

4 Spot some Quink on the origin using a small clean loop of nichrome wire.

5 Dry the ink mark, add a second drop to concentrate and then dry again.

6 Spot the other two chromatoplates in the same way using *(a)* brown ink and *(b)* an equal mixture of brown ink and black Quink.

7 Develop the plates by standing them in the beaker of solvent, taking care that the origin is not immersed.

8 Run until the solvent front reaches the top of the plates, then remove and dry.

Sequel

1 Examine the chromatograms of the individual inks and identify as many coloured ingredients as possible. Are any of the ingredients common to both inks?

2 Can you recognize the individual ingredients of the inks on the mixed chromatogram?

8.4 EXTRACTION OF FLUORESCING AGENTS FROM DETERGENT

(TLC and location by fluorescence)

Required

Synthetic detergent powder; ethanol (industrial methylated spirit); ammonia solution (0.88); Silica Gel G (chromatographic grade); methylbenzene; ethyl ethanoate; pyridine*

Ultraviolet lamp

Procedure

1 Add 50 cm^3 of ethanol and five drops of 0.88 ammonia solution to 1 g of synthetic detergent powder in a boiling tube.

2 Gently shake the mixture, preferably warming gently under a running hot water tap. Allow to cool.

3 Prepare two chromatoplates as in the previous experiment using a 2:1 slurry of water and silica gel. Dry the plates at 110 °C for at least an hour.

4 Using a melting-point tube place a small spot of the cold detergent extract about 1 cm from the end of each chromatoplate.

5 Place the two prepared plates V-wise, spotted end down, into a beaker containing an equal volume mixture of methylbenzene and ethyl ethanoate to a depth of about 0.5 cm.

6 Add two drops of pyridine to the solvent with a small pipette and cover the beaker with a watch glass.

7 When the solvent reaches the top of the plates remove them carefully and lay on filter papers, treated side uppermost, to dry.

8 View the dried plates under an ultraviolet lamp and note the position of any fluorescing material.

Sequel

Does there appear to be more than one type of fluorescing agent present in the detergent? If so why is this necessary?

***Pyridine is very toxic and should be handled with care.**

8.5 TRIAL SEPARATION OF MALACHITE GREEN AND METHYLENE BLUE

(wedge and sandwich chromatography)

Required

Malachite green; methylene blue; ethanol (industrial methylated spirit); silica gel and aluminium(III) oxide (both chromatographic quality); methanol

Procedure

1 Prepare three wedges of silica gel and three of aluminium(III) oxide (fig 8.10).

2 Dissolve 2 mg of malachite green and methylene blue in 2 cm^3 of ethanol and spot a little of the mixture on the front edge of each wedge.

3 Develop a wedge of each adsorbent by pouring a little methanol into the dish.

4 Repeat with the other two pairs of wedges using ethanol and distilled water as the developing solvents.

Sequel

1 Look at the six chromatograms produced. Suggest an effective technique for separating the given mixture by column chromatography.

2 Repeat the experiment using the sandwich technique (fig 8.10). Compare the results from the two techniques.

8.6 PURIFICATION OF ANTHRACENE

(locating agent with column chromatography)

Required

Anthracene (technical grade); aluminium(III) oxide; hexane (or petroleum ether, 80-100 °C b.p. range)

Procedure

1 Using a semi-micro assembly (fig 8.3) prepare a column of aluminium(III) oxide by the wet method (fig 8.5a), making the slurry up with hexane or petroleum ether.

2 Dissolve 0.1 g of crude anthracene in 10 cm^3 of hexane and run through the column.

3 Develop with about 10 cm^3 of hexane and then examine the column under an ultraviolet lamp.

4 Two fluorescent bands should be seen — an upper blue line produced by carbazole and a lower wide band of a characteristic blue-violet fluorescence produced by the pure anthracene. In addition a middle yellow band of naphthacene is usually present which does not fluoresce under ultraviolet light.

Sequel

1 Continue to develop the column using hexane or petroleum ether until the lower anthracene band has been completely eluted. (Check with the ultraviolet lamp.)

2 If time allows recover pure anthracene (m.p. 216 °C) from the eluate by distilling off the solvent at reduced pressure.

8.7 IDENTIFICATION OF FOOD DYES BY ASCENDING CHROMATOGRAPHY

(multiple ascending chromatography)

Required

Food colours (Boots' Besyet); unknown mixture of food colours; ethanol (industrial methylated spirit)

Procedure

1 Prepare a sheet of chromatographic paper for a multiple run (fig 8.13).

2 Spot the coloured dyes along the origin of the paper using a clean wire loop, reserving one place for a drop of the unknown mixture.

3 Roll the paper into a cylinder, clip the sides together and develop with a 1:4 solution of ethanol and water in a closed gas jar.

4 Remove and dry the chromatogram when the solvent front has reached the upper edge of the slots in the paper.

Sequel

Rule pencil lines horizontally across the paper through the centre of each coloured spot of the resolved mixture. Observe where these lines cross the spots on the other chromatograms and suggest which of the colours were present in the mixture.

8.8 ANALYSIS OF A MIXTURE OF INDICATORS

(multiple descending chromatography)

Required

Congo red; phenol red; bromophenol blue; methyl orange; unknown mixture of indicators; ethanol (industrial methylated spirit); saturated ammonium sulphate solution

Procedure

1 Prepare a paper sheet for downward chromatography (fig 8.14).

2 Prepare spots of each of the four indicators along the origin, and then a spot of the unknown mixture.

3 Half fill the solvent trough with a mixture of water/ethanol/saturated aqueous ammonium sulphate in a volume ration of 5:2:1.

4 Run the chromatogram until the solvent front approaches the lower edge of the paper, then remove and dry.

Sequel

Rule pencil lines horizontally across the paper through each indicator spot on the chromatogram of the mixture. Observe where these lines cross the other chromatograms and suggest which of the indicators were present in the mixture.

8.9 SEPARATION OF AMINOACIDS IN HUMAN HAIR HYDROLYSATE

(two-dimensional separation and locating reagent)

Required

Tetrachloromethane; mixture of dilute hydrochloric acid and water (1:1); propanone; benzene; ethanol (industrial methylated spirit); ammonia solution (0.88); butan-1-ol; ethanoic acid; ninhydrin solution (0.2 g in 100 cm^3 of propanone)

Tuft of human hair (or cat fur); wool

Broken porcelain or boiling chips

Procedure

1 Rinse a small tuft of human hair with tetrachloromethane and allow to dry.

2 Heat the hair in a boiling tube with about 10 cm^3 of dilute hydrochloric acid (1:1 with water) and a few pieces of broken porcelain using a cold finger condenser to reflux the boiling liquid.

3 At the end of an hour or so the hair should have been degraded forming a brownish liquid.

4 Filter the liquid on a Hirsch funnel and then reduce the volume to about 1 cm^3 by evaporation in an open dish over a water bath.

5 Rinse the sides of the dish with a few cm^3 of an equal volume mixture of propanone and benzene and then evaporate to dryness.

6 Take up the residue with a drop or two of ethanol.

7 Mark an origin near one corner of the sheet of chromatograph paper as shown in fig 8.16a.

8 Spot the mixture on the origin, then dry and repeat. Run the first chromatogram using a mixture of water/ethanol/ammonia in a ratio 1:8:1. If a tank is not available roll the paper into a cylinder and use a gas jar (fig 8.17).

9 Allow the solvent front to overrun the opposite edge, then remove the paper and dry.

10 Turn the paper through 90° and make the second run using a mixture of butan-1-ol/ethanoic acid/water in a ratio of 4:1:2.

11 Again allow the solvent to overrun the edge and then remove the paper and dry.

12 Spray or dip the paper with the solution of ninhydrin in propanone and warm in an oven for a few minutes at 100 °C.

13 Note the brown, purple and yellowish spots which indicate the presence of aminoacids and peptides, and ring their position because they often fade quite quickly in daylight.

Sequel

Repeat the experiment using undyed wool and compare the two chromatograms.

8.10 ANALYSIS OF INKS

(disc chromatography)

Required

Selection of coloured inks;
ethanol (industrial methylated spirit)

Procedure

1 Prepare a number of paper discs using one of the techniques illustrated in fig 8.18a.

2 Spot the centre of each disc several times with a selected ink, drying between each application (or soak the wick in the ink and dry).

3 Run the chromatograms using a 1:4 mixture of ethanol and water.

4 When the solvent reaches the edge of the disc, stop the run and dry.

Sequel

1 Cut the best sections from the discs and determine the R_f values of the coloured components.

2 Make a list of the inks and their probable coloured ingredients.

8.11 PURIFICATION OF CRUDE SUGAR

(ion exchange chromatography)

Required

Zerolit FF (IP) (Cl^- form)*; Zerolit 226 (H^+ form)*; sodium hydroxide solution (1 M)

Equal mixture of brown and granulated sugar

Procedure

1 Fill a measuring cylinder to the 20 cm³ mark with Zerolit FF and cover with 2.5 cm of water. Gently tap the cylinder wall and add more resin if necessary to top up to the mark again.

2 Pour the resin into a 1.25 cm diameter column (fig 8.19a) and regenerate with 40 cm³ 1 M sodium hydroxide solution. (This should take about 10 minutes to pass through the column.)

3 When the sodium hydroxide has almost drained to bed level, rinse the column with 200 cm³ of distilled water (40 minutes).

4 Empty the resin from the column into a beaker and stir in 10 cm³ of Zerolit 226 measured out as before. After mixing return the resin to the column and rinse with distilled water several times. Take care not to allow the surface of the resin to become uncovered.

5 Dissolve 40 g of the sugar mixture in 200 cm³ of water and run through the column at a rate of about 4 cm³ a minute.

Sequel

1 Compare the colour of the eluate from the column with the original solution. Does this indicate that the colouring matter was ionic?

2 Evaporate a measured volume of the eluate to dryness using a water bath and weigh the residue. Compare the concentration of the eluted sugar solution with that of the original mixture.

8.12 CONVERSION OF TRISODIUM(I) CITRATE TO 2-HYDROXYPROPANE-1,2,3-TRICARBOXYLIC ACID (CITRIC ACID)

(ion exchange chromatography)

Required

Zerolit 226 (H^+form)*; trisodium(I) citrate;
hydrochloric acid (10%)

Procedure

1 Measure out 20 cm³ of Zerolit 226 resin and empty into a 1.25 cm diameter column as in the previous experiment.

2 Rinse the column with distilled water several times.

3 Make a solution of 50 mg of trisodium(I) citrate in 100 cm³ of water and pass through the prepared column at a rate of about 4 cm³ a minute. When the solution has all passed through rinse the column with 50 cm³ of distilled water.

*These resins are obtainable from the Permutit Co. Ltd, Gunnersbury Avenue, London W4 or from Hopkin & Williams, Freshwater Road, Chadwell Heath, Essex.

4 Concentrate the eluted solution and washings to a small bulk by evaporation over a water bath. Cool and allow to crystallize.

5 Filter off the crystals using a Hirsch funnel and dry with filter paper.

Sequel

1 Confirm the identity of the substance by determining its melting point (2-hydroxypropane-1,2,3-tricarboxylic acid m.p. 100 °C).

2 Regenerate the column by washing with 10% hydrochloric acid and then distilled water. What substance would you expect to find in the wash water?

3 Suggest a means of separating the components of lemonade powder which contains sucrose, 2-hydroxypropane-1,2,3-tricarboxylic acid, and colouring matter.

8.13 DEIONIZATION OF A SOLUTION

Required

Zerolite 225 (Na^+ form)*; Zerolit FF (IP) (Cl^- form)*; hydrochloric acid(1M); sodium hydroxide solution(1M); copper(II) sulphate(VI); dipotassium dichromate(VI); aqueous solution (1%) of lead(II) nitrate(V); aqueous solution (1%) of potassium iodide

Procedure

1 Make up a column of Zerolit 225 in a 1.25 cm glass tube using about 25 cm³ of resin as in experiment 8.11.

2 Wash the column with about 40 cm³ of 1 M hydrochloric acid to regenerate the H^+ form and then rinse with distilled water.

3 Similarly make up a second column of Zerolit FF and regenerate with 1 M sodium hydroxide solution.

4 Make up dilute solutions of copper(II) sulphate(VI) and dipotassium dichromate(VI) in distilled water, adding only sufficient solid to produce a definite coloration.

5 Mix the two solutions and pour about 50 cm³ of the mixture into the Zerolit column, which should remove the potassium and coloured copper(II) ions, leaving an orange eluate.

6 Finally pour this orange solution containing sulphate(VI) and coloured dichromate(VI) ions through the Zerolit FF column and note that the eluate is completely decolorized.

Formulae

Zerolit 225

$$2R(H^+) + CuSO_4 \rightleftharpoons R_2(Cu^{2+}) + H_2SO_4$$

$$2R(H^+) + K_2Cr_2O_7 \rightleftharpoons 2R(K^+) + H_2Cr_2O_7$$

Zerolit FF

$$2R(OH^-) + H_2SO_4 \rightleftharpoons R_2(SO_4^{2-}) + 2H_2O$$

$$2R(OH^-) + H_2Cr_2O_7 \rightleftharpoons R_2(Cr_2O_7^{2-}) + 2H_2O$$

*These resins are obtainable from the Permutit Co. Ltd, Gunnersbury Avenue, London W4 or from Hopkin & Williams, Freshwater Road, Chadwell Heath, Essex.

Sequel

1 One-third fill a boiling tube with distilled water and add a few drops each of lead(II) nitrate(V) solution and potassium iodide solution, producing a yellow precipitate of lead(II) iodide.

2 Shake with about 10 cm³ of Zerolit 225 resin and notice that the colour disappears. What has taken place?

8.14 EXAMINATION OF CIGARETTE LIGHTER FUEL

(gas–liquid chromatography)

Required

Capsule of cigarette lighter fuel

Stop-clock

Procedure

1 Set up the assembly shown in fig 8.21.

2 Heat the water in the beaker to a temperature of 70 °C and maintain at this temperature throughout the experiment.

3 The bath should be occasionally gently agitated either with a stirrer as illustrated or by gently twisting the wooden lid, using the coil as a stirrer.

4 Turn on the gas and after a few moments light the jet. Adjust the gas to give a small non-luminous flame about 5 mm high.

5 Pierce the lighter fuel capsule with a syringe needle and suck up a few drops of the liquid.

6 Remove the syringe and push the needle through the plastic tubing as shown, taking care not to inject any liquid.

7 Start a stop clock and as the hand reaches the zero mark, inject a drop of the lighter fuel so that it falls on the column surface.

8 Watch the flame carefully; the arrival of the hydrocarbon components at the jet will be signalled by a sudden luminosity of the flame which after a short time will become non-luminous again. The time of arrival and duration of each component should be noted.

Sequel

1 Tabulate your results, listing the appearance of the fractions in chronological order.

2 Inject a number of hydrocarbon liquids of low boiling point individually into the apparatus. Note the time taken for them to appear at the jet and the temperature of the water bath (e.g. petroleum ethers of various boiling point ranges, pentane, hexane, heptane, and cyclohexane). Use the data you obtain to suggest some of the ingredients of lighter fuel.

8.15 SEPARATION OF DYES

(electrophoresis)

Required

Aqueous solutions (1%) of tartrazine, rhodamine, and fluorescein; ethanoate buffer pH 10.0

Whatman 3 MM chromatography paper strip

Procedure

1 Fill the buffer compartments of the cell (or saturate the sponge blocks according to type) with ethanoate buffer pH 10.0.

2 Cut two strips of chromatography paper long enough to stretch between the cell electrodes and lightly draw a pencil line across the centre as the origin.

3 Mark one end of the paper with a cross to indicate the anode junction.

4 Soak the paper strips in the buffer solution, allow to drain, then blot off the excess liquid and fix the strips in position in the tank with the cross at the anode end.

5 Replace the lid on the tank and switch on the current, which should be about 350 V d.c. for a strip 25 cm long, and allow to equilibrate for five minutes.

6 Switch off the current supply. After waiting for one minute remove the tank lid and pipette a small spot of tartrazine on the centre of each origin. Then pipette similar amounts of rhodamine and fluorescein in the centre of the previous spot.

7 Replace the lid immediately and switch on the power.

8 Allow the electrophoresis to continue until a good separation has been achieved.

9 Switch off the power, wait for one minute before opening the tank and then lift out the strips and dry them in an oven or by using a hair-dryer.

10 What can you say about the overall charge on the dyes?

Sequel

Evaporate a little of the green liquid from a tin of processed peas to concentrate. Place a spot on the origin of a piece of paper strip prepared for electrophoresis as before. Alongside the first spot pipette a small drop of green food dye (Boots' Besyet). After carrying out electrophoresis for half an hour examine the dye samples and comment on the results. (A suitable, reasonably priced apparatus for carrying out this experiment is the Instrolec 100 Electrophoresis Tank.)

8.16 SEPARATION OF AMINOACIDS IN LEMON JUICE

(two dimensional separation:
electrophoresis and chromatography)

Required

Lemon juice; ethanoic acid; pyridine;
ethanol (industrial methylated spirit; ammonia solution (0.88); ninhydrin in propanone (100 mg in 50 cm³)

Procedure

1 Draw a pencil line across the centre of a filter paper strip and mark a single origin on it 2 cm away from one edge (fig 8.24).

2 Spot the origin with fresh lemon juice, dry, make a second application of juice, and redry. Repeat this operation five times.

3 Make up the buffer solution by adding 0.8 cm³ of ethanoic acid to 10 cm³ of pyridine and making up to 250 cm³ with distilled water. Check the pH which should be just over 6.0 using an indicator paper or pH meter.

4 Soak the paper in the buffer as in the previous experiment and fix in position after filling the buffer compartments of the cell.

5 Run for one hour at 350 V d.c. then switch off the current and dry the sheet quickly with a current of unheated air.

6 Roll the dry sheet into a cylinder with the electrophoretogram along the lower edge and clip the upright edges together.

7 Place the cylinder in a gas jar containing a mixture of ethanol/water/0.88 ammonia solution in the proportions of 8:1:1 and run the chromatogram until the solvent front reaches the upper edge of the paper.

8 Dry the two-way chromatogram and spray or dip into the ninhydrin solution.

9 Heat in an oven for a few moments at 80 °C, when the aminoacids will appear as blue, brown or pink spots.

10 Ring each spot with a pencil mark as rapid fading occurs in daylight.

Sequel

Repeat the experiment using other fruit juices and compare your results.

8.17 DESALTING SKIMMED MILK BY GEL FILTRATION

Required

Skimmed milk diluted 1:1 with distilled water (Marvel powdered milk reconstituted with twice the recommended volume of water is satisfactory)

Sephadex gel filtration kit;
Sephadex G 25 ion exchange gel;
Benedict's reagent; nitric(V) acid (1 M);
silver(I) nitrate(V) solution

5 cm^3 syringe; 5 cm^3 and 1 cm^3 pipettes

Procedure

Preliminary note: the outline procedure for setting up the column is given; more detailed instructions are included in the booklet *Sephadex — Gel Filtration in Theory and Practice* supplied by the Pharmacia Company* who also supply a gel filtration kit for school use. The gel should be swollen before use by soaking about 5 g in distilled water for 1-2 days at room temperature or for two hours over a boiling water bath.

1 Set the column up vertically and using a syringe attached to the outlet, force in enough distilled water to ensure that the bed support is covered to a depth of about 1 cm. Ensure that no bubbles of air are trapped in the support by passing the eluent backwards and forwards two or three times and then close the outlet and remove the syringe.

2 Make a thin slurry of the fully swollen ion exchanger in distilled water and pour steadily into the column, avoiding bubble formation. Allow the gel to stabilize for 5–10 minutes and then top up the column with distilled water and attach the eluent reservoir.

*Pharmacia (Great Britain) Ltd, Paramount House, 75 Uxbridge Road, London W5 5SS.

3 Open the column outlet and allow a few cm^3 of distilled water from the reservoir to run through the column into a conical flask at the flow rate to be used during the experiment. The flow rate can be varied if required by raising or lowering the eluent reservoir and thus altering the pressure head (fig 8.22). A pressure head of 30 cm gives a satisfactory flow rate for this experiment of 8-10 cm^3/hour.

4 Close the column outlet, disconnect the eluent reservoir, and remove excess distilled water from the bed surface using a pipette. Using a clean pipette transfer 1 cm^3 of the diluted skimmed milk to the drained surface and open the column outlet. When the sample has been sucked into the bed, top up the column with distilled water and replace the eluent reservoir.

5 Use labelled test tubes to collect eluted fractions every 15 minutes (2–2.5 cm^3), over a period of 2½ hours, and note the appearance of each fraction.

6 Divide each fraction into two parts and treat as follows:

Part A; boil with a little Benedict's reagent for 30 seconds. Red-brown precipitate indicates the presence of lactose.

Part B: boil with 0.1 cm^3 of dilute nitric (V) acid. Centrifuge or filter off any precipitate formed and then add 0.2 cm^3 of silver(I) nitrate(V) solution. Note in which sample the chloride ion first appears.

7 After use, or if required for further analysis, the column should be washed with at least one column volume of 1% sodium hydroxide solution at about 60 °C.

Note: in the wet state Sephadex ion exchangers are stable for several months if stored in a buffered suspension containing an antimicrobial agent. Microbial growth interferes with the performance of the gel and may obstruct flow through the bed. A 0.05% solution of trichlorobutanol in buffer solution (pH 5) is suitable for this purpose.

ACKNOWLEDGEMENTS

Thanks are due to the following who have kindly permitted the reproduction of copyright photographs: page 2, A. Gallenkamp & Co. Ltd; page 13, Griffin & George Ltd; page 14, May & Baker Ltd; page 26, Hylton Warner; pages 36, 49, Shell Petroleum Co. Ltd; and to the following for permission to base the diagrams shown on material from their publications: figure 1.4, A. Gallenkamp & Co. Ltd; 3.7, Shell International Petroleum Co. Ltd; 5.3, 6.4, 6.5, 6.8, 6.9, 6.10, 7.2, 7.3, Quickfit & Quartz Ltd; 6.2, Griffin & George Ltd; 8.6, Shandon Southern Instruments Ltd; 8.24, Pharmacia (Great Britain) Ltd.

Thanks are also due to Unilever Ltd for permission to base experiment 8.4 on their Laboratory Experiment No. 11, and to Pharmacia (Great Britain) Ltd for experiment 8.17 from their booklet *Sephadex – Gel filtration in theory and practice.*

APPENDIX Chemical nomenclature

Recommended name	*Common current name*
anthracene-9,10-dione	9,10-anthraquinone
benzenecarbaldehyde	benzaldehyde
benzenecarboxamide	benzamide
benzenecarboxylic acid	benzoic acid
benzene-1,4-diamine	*p*-phenylenediamine
benzene-1,2-dicarboxylic anhydride	phthalic anhydride
benzene-1,3-diol	resorcinol
buta-1,3-diene	butadiene
butanedioic acid	succinic acid
butan-1-ol	n-butyl alcohol
butyl ethanoate	butyl acetate
carbamide	urea
cellulose ethanoate	cellulose acetate
chloromethane	methyl chloride
chloromethoxymethane	chloromethyl ether
1,4-dichlorobenzene	*p*-dichlorobenzene
1,3-dinitrobenzene	*m*-dinitrobenzene
ethanoic acid	acetic acid
ethoxyethane	diethyl ether
ethyl ethanoate	ethyl acetate
2-hydroxybenzoic acid	salicylic acid
2-hydroxypropane-1,2,3-tricarboxylic acid	citric acid
methanal	formaldehyde
methanol	methyl alcohol
methoxymethane	dimethyl ether
methylbenzene	toluene
2-methylphenylamine	*o*-toluidine
4-methylphenylamine	*p*-toluidine
2-methylpropan-2-ol	tert-butyl alcohol
naphthalen-2-ol	*β*-naphthol
2-nitrophenylamine	*o*-nitroaniline
4-nitrophenylamine	*p*-nitroaniline
pentyl ethanoate	amyl acetate
phenylamine	aniline
N-phenylethanamide	acetanilide
phenylethene	styrene
poly(ethene)	polythene
poly(phenylethene)	polystyrene
potassium(I) sodium(I) 2,3-dihydroxybutanedioate	Rochelle salt
propan-1-ol	n-propanol
propanone	acetone
tetrachloromethane	carbon tetrachloride
tetraphosphorus decaoxide	phosphorus pentoxide
trichloromethane	chloroform
triiodomethane	iodoform